Lecture Notes for Human Anatomy and Physiology

Second Edition

Ann M. Findley
Amy G. Ouchley

Department of Biology
University of Louisiana at Monroe

Copyright © 1998, 1999 by Kendall/Hunt Publishing Company

ISBN 978-0-7575-2406-6

All rights reserved. No part of this publication may be reproduced, stored in a retrieval system, or transmitted, in any form or by any means, electronic, mechanical, photocopying, recording, or otherwise, without the prior written permission of the copyright owner.

Printed in the United States of America

Table of Contents

	page
Chapter 1 - Overview of Human Anatomy & Physiology	1
Chapter 2 - Chemical Basis of Life	5
Chapter 3 - Cell Structure and Function	11
Chapter 4 - Histology	17
Chapter 5 - The Integumentary System	23
Chapter 6 - The Skeletal System	27
Chapter 7 - The Skeletal System: Important "Facts" to Know & Apply	31
Chapter 8 - Articulations and Movements	33
Chapter 9 - The Muscular System: Histology/Physiology	35
Chapter 10 - The Muscular System: Gross Anatomy	43
Chapter 11 - Functional Organization of Nervous Tissue	45
Chapter 12 - The Spinal Cord and Spinal Nerves	55
Chapter 13 - The Brain and Cranial Nerves	57
Chapter 14 - Integration of Nervous System Function	63
Chapter 15 - The Special Senses	67
Chapter 16 - The Autonomic Nervous System	73
Chapter 17 - Functional Organization of the Endocrine System	75
Chapter 18 - The Endocrine Glands	77
Chapter 19 - The Blood	83
Chapter 20 - The Heart	89
Chapter 21 - Peripheral Circulation & Regulation	99
Chapter 22 - The Lymphatic System & Immunity	107

Chapter 23 - The Respiratory System ... 113

Chapter 24 - The Digestive System ... 119

Chapter 25 - Nutrition, Metabolism and Temperature Regulation ... 125

Chapter 26 - The Urinary System ... 129

Chapter 27 - Fluid, Electrolyte and Acid-Base Balance ... 135

Chapter 28 - The Reproductive System ... 139

Appendix - Review Sheets ... 145

Chapter 1 - Overview of Human Anatomy & Physiology

Anatomy - Study of structure.
Cytology -   **Histology** -
Gross anatomy-   **Systemic anatomy** -

Physiology - Study of function.
Cell physiology -   **Neurophysiology** -
Human physiology-

Structure/Function Relationships - provides the basis for understanding disease and how the body maintains a **dynamic** state of **homeostasis**.

Seven Structural Levels of the Body: (*Note the increasing complexity)

Chemical → Organelle → Cell → Tissue → Organ → Organ System → Organism

1. **Chemical** - this level of organization involves interactions between atoms and their combinations into molecules. (Chapter 2)
2. **Organelle** - a small structure contained within a cell that performs one or more specific functions. (Chapter 3)
3. **Cell** - basic living unit of all plants and animals that while differing in details of structure/function share many common characteristics. (Chapter 3)
4. **Tissue** - group of cells with similar structure/function. (Chapter 4)
5. **Organ** - composed of two or more types of tissues that perform one or more common functions.
6. **Organ System** - group of organs classified as a unit because of a common function or set of functions.
7. **Organism** - any living thing considered as a whole. The human organism is the complex sum of organ systems that are mutually dependent upon each other.

Organ System Overview

1. **Integumentary:** protection, regulation of body temperature (T), prevention of H_2O loss, production of vitamin D precursor (skin, hair, nails, sweat/oil glands)

2. **Skeletal:** protection, support, participation in body movements, blood cell production, mineral storage (bones, cartilage, joints).

3. **Muscular:** participation in body movement, posture maintenance, heat generation.

4. **Nervous:** coordination/regulation: sensation, controls movement, integrated control of physiological/intellectual functions (brain, spinal cord, nerves, and sensory receptors).

5. **Endocrine:** coordination/regulation: metabolic regulation, reproduction, etc. (Includes cells/glands that secrete **hormones** = chemical messengers.)

6. **Cardiovascular System:** transports nutrients, waste products, gases, and hormones; plays role in immune response and regulation of body T (heart, blood vessels, blood).

7. **Lymphatic:** removes foreign substances from the blood/lymph; combats disease, maintains tissue fluid balance; participates in fat absorption (lymph vessels, nodes, organs).

8. **Respiratory:** gas (O_2/CO_2) exchange between blood (capillaries) & alveoli (lungs), regulates blood pH (lungs and respiratory passages).

9. **Digestive:** performs mechanical & chemical processes of digestion, absorption of nutrients, and elimination of wastes (tube from mouth → anus + accessory organs).

10. **Urinary:** removes waste products from the blood, regulates fluid/electrolyte balance (kidneys, urinary bladder, ureters, urethra).

11. **Reproductive:** performs process of reproduction, controls sexual functions & behaviors (gonads, ducts, accessory structures, external genitalia).

Characteristics of Life

1. **Organization:** Living things are highly organized. Disruption of organization can result in loss of function or death.
2. **Metabolism:** the ability to use energy to perform vital functions.
3. **Responsiveness:** An organism is responsive if it can sense changes in its internal & external environment and make adjustments that help maintain life.
4. **Growth and Development:** Growth is the ability to increase in size or number. Development includes changes an organism undergoes from fertilization to death.
5. **Reproduction:** Formation of new cells or new organisms, necessary for the continuation of life.

HOMEOSTASIS:

- The existence or maintenance of a **relatively constant or stable environment** within the body.
- State of **dynamic equilibrium**.
- Maintenance of a variable around an ideal normal value (= **set point**). The value of the variable fluctuates around the set point, establishing a normal range of values.
- Several organ systems (1° nervous & endocrine) help control the internal environment.
- Disruption of homeostasis results in disease and possibly death.

Example of homeostatic mechanism: Body Temperature Regulation
- Body's thermostat is located in the hypothalamus of the brain.
- With ↑ T, sweat glands produce perspiration (= evaporative cooling) & superficial blood vessels dilate (= radiative cooling).
- With ↓ T, skeletal muscle contraction produces shivering (generates heat) & superficial blood vessels constrict to shunt blood away from surface (↓ heat loss).

Stimulus (stressor) → Receptor → Control Center → Effector → Response

Negative Feedback Control: Reversing a change to keep a variable within a **normal** range.
Positive Feedback Control: Increases deviations from normal; may create "vicious cycles", but can be constructive.
Examples:

Read & Know: Clinical Focus: Anatomical Imaging
Body plan - bilateral symmetry
Anatomical position:
Directional Terms for Humans (Table 1.1, Fig. 1.10)
Body Planes: sagittal (longitudinal); transverse/horizontal (cross-section); frontal/coronal
Body Regions (Fig. 1.11)
Body Cavities: dorsal
   thoracic - pleural cavities, mediastinum, pericardial cavity
      ventral ---------------------- diaphragm --------------------------------
   abdominopelvic -abdominal cavity, pelvic cavity

Serous Membranes: Parietal -   Visceral-

　　　　　　　　　　Pleural -   Peritoneal -

Anatomical Comparative Terms:

Superior -   Inferior -

Cephalic -   Caudal -

Anterior -   Posterior -

Ventral -   Dorsal -

Proximal -   Distal -

Lateral -   Medial –

Superficial -   Deep -

Chapter 2 - Chemical Basis of Life

matter - occupies space and has mass

elements - substances that can't be broken down into simpler substances by ordinary chemical reactions
- 109 total; 92 natural & 17 'artificial'

atom - building blocks of elements
    neutron - no electrical charge, 1 amu ⎫ found within
    proton - +1 electrical charge, 1 amu ⎬ nucleus
    electron - -1 electrical charge, 0 amu ⎭

atoms are electrically neutral: #p = #e

Atomic number = Z = #p (also = #e)

Mass number = #p + #n

Isotopes - atoms with the same #p & e **but** different #n   e.g., $^1H, ^2H, ^3H$

    radioactive isotopes - unstable nuclei that undergo decay
    - used in clinical imaging techniques, ^{14}C dating of biological materials

molecules - 2 or more atoms joined together by chemical bonds

chemical bonds - atoms in a molecule **transfer or share** e⁻ in their outermost shell

1- **ionic bonds** - opposite-charged **ions** are attracted to one another   e.g., Na^+Cl^-

    ions - atoms that have gained or lost an electron

        cation - loses e⁻, becomes + charged, metals
        anion - **gains** e⁻, becomes - charged, nonmetals

2- **covalent bonds** - sharing of e⁻ pairs between atoms in a molecule

    nonpolar covalent - **equal** sharing   e.g., H_2, O_2, N_2

    polar covalent - **unequal** sharing, molecule has + and - regions (e.g., H_2O)

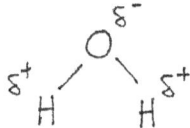

3- **hydrogen bonds** - weak attraction between H already covalently-bonded to some element and O or N on another molecule (**intermolecular H bond**) or at another place on the same molecule (**intramolecular H bond**)

Chemical reactions -   reactants → products;   ≡ reversible reaction

Reaction types:

1) **synthesis reaction** -   A + B → AB   anabolic, require energy
2) **decomposition reaction** -   AB → A + B   catabolic, releases energy
3) **exchange reaction** -   AB + CD → AD + CB
        e.g., HCl + NaOH → NaCl + H_2O neutralization reaction

  dehydration reaction - H_2O is a product

  hydrolysis reaction - H_2O is a reactant

4) **oxidation-reduction reaction** - transfer of e^-

    oxidation = loss of e^- **reduction** = gain of e^-

Reaction rates: affected by nature & concentration of reactant(s)
    ↑ **rate** ⇒ ↑ **concentration,** ↑ **T, presence of catalyst (enzyme)**
    catalyst - substance that increases rate at which reaction proceeds
        without being permanently changed or depleted.
    enzymes - **protein** molecules that act as biological catalysts

Energy - capacity to do work

  potential energy - stored energy, due to stationary position
  kinetic energy - energy caused by movement, actually does work

  exergonic reactions - release energy; bonds broken; primarily catabolic
  endergonic reactions - require energy input; bonds formed; primarily anabolic

activation energy (E_{act}) - amount of energy required to start a reaction
   – "energy barrier" that must be overcome
   – enzymes increase reaction rates by decreasing the activation energy

Water - 60-80% of the volume of most cells
— **intra**molecular vs. **inter**molecular bonding interactions

+ liquid @ room temperature - important as a transport medium, solvent
+ high specific heat - resists ΔT, helps body T regulation
+ protection - good lubricant, reduces friction/wear & tear
+ mixes well with polar molecules - "biological solvent"
    solution - liquid (solvent) with dissolved substances (solute)
    suspension - liquid with undissolved substances, settle out without movement
        – e.g., RBC in plasma
    colloid - liquid with undissolved substances that won't settle out
        – e.g., plasma proteins
+ participates in biological reactions as a reactant/product

    glucose + fructose → sucrose + H_2O
    sucrose + H_2O → glucose + fructose

Acids and Bases

acids - H^+ donor   **bases** - H^+ acceptor, many bases release OH^-

pH = - log [H^+]   pH scale 0 → 14

acidic solutions -	[H^+] > [OH^-]	pH less than 7
neutral solutions -	[H^+] = [OH^-]	pH = 7
basic solutions -	[H^+] < [OH^-]	pH greater than 7

Buffers - pair of chemicals that resist pH change
    e.g., H_2CO_3/HCO_3^- - carbonic acid/bicarbonate ion; 1° plasma buffer
        - weak acid/salt of a weak acid
        - acts as base, neutralizes excess acid (HCO_3^-)
        - acts as acid, neutralizes excess base (H_2CO_3)

$$CO_2 + H_2O \overset{CA}{\Leftrightarrow} H_2CO_3 \Leftrightarrow H^+ + HCO_3^-$$ (CA = carbonic anhydrase)

CARBOHYDRATES - sugars, polar molecules (water soluble), $C_nH_{2n}O_n$

Monosaccharides - simple sugars C_3 - C_7
    C_5 = pentoses - ribose, deoxyribose - nucleic acid sugars
    C_6 = hexoses - <u>glucose, galactose</u>, <u>fructose</u> - $C_6H_{12}O_6$ isomers
                   aldoses   ketose

Disaccharides - two simple sugars joined by glycosidic (polar covalent) bond

    sucrose = glucose + fructose   (table sugar)
    maltose = glucose + glucose   (malt sugar)
    lactose = glucose + galactose   (milk sugar)

Polysaccharides - complex CHO, many simple sugars bound together

    glycogen = animal starch, primarily found in liver & skeletal muscle
    starch = plant storage material
    cellulose = plant structural CHO (in CW); nondigestible by humans

LIPIDS - Fats, nonpolar molecules (water insoluble)

Neutral fats - glycerol + 3 FA = triacylglycerol (triglyceride)

    saturated fat - only C-C, solid, 1° animals products, "fats"
    unsaturated fat - 1 or more C=C, liquid, 1° plant origin, "oils"

Phospholipids - glycerol, 2FA, phosphate group, polar molecule

  – found in CM
  – amphipathic - have polar + nonpolar regions

    polar head group   HC tail, nonpolar
    hydrophilic   hydrophobic

Steroids - 4 fused C rings

  – important as hormones, vitamin D, bile

FA derivatives - prostaglandins (PGs), local hormones

PROTEINS - C,H,O,N,S; made up of aa (20 aa - 11 nonessential/9 essential)

$H_2N - \overset{H}{\underset{R}{C}} - COOH$

- important as hormones, enzymes, Abs, receptors, transporters
- aa joined together by peptide bonds (polar covalent)
- ability to function depends on native conformation (3D shape)

1° structure - linear aa sequence
2° structure - α helix (coil, spring), β pleated sheet (zig-zag, accordian), H bonds
3° structure - 3D conformation/shape, fibrous (ropelike) vs. globular (spherical)
4° structure - subunit association, interaction between polypeptide chains

enzymes - increase reaction rate by decreasing E_{act}, end in -ase, specificity

active site = substrate binding site
lock-and-key mechanism - specificity of fit/action
competitive vs. noncompetitive (allosteric = other site) inhibition
cofactor - ions that assist enzyme function
coenzyme - vitamins that assist enzyme function

denaturation - protein unravels, loses 2° & 3° structure but 1° structure remains
– proteins are sensitive to changes in T & pH
- shape changes → loss of function

NUCLEIC ACIDS - made up of nucleotides = sugar, phosphate, N-base

sugars = ribose, deoxyribose
bases = guanine (G), adenine (A) = purines
cytosine (C), thymine (T), uracil (U) = pyrimidines

RNA - ribose, phosphate, G,A,C,U; single-stranded, α helix
– 3 types = rRNA, tRNA, mRNA

DNA - deoxyribose, phosphate, G,A,C,T; double-stranded, α helix

```
P           P
 \         /
  S - A····T - S         S-P, base-S → polar covalent bonds
 /         \             base ||||base → H bond
P           P            G····C,  A····T
 \         /
  S - G····C - S
 /         \
P           P
```

ATP - adenosine triphosphate = A, ribose, 3 phosphate groups

```
A
|
ribose - P ~ P ~ P     high-energy bond
```

"energy currency", produced primarily within mitochondria

Chapter 3 - Cell Structure and Function

cell - fundamental unit of life
  – bordered by plasma or cell membrane (PM/CM)

extracellular - outside the CM (ECF)
intracellular - inside the CM, within the cell (ICF)

CM - encloses/supports the cell; selectively permeable
  – recognition and communication functions
  – consists primarily of a PL bilayer (lipid "sea", "molecular sandwich")
  – cholesterol found in animal CM, ↓ CM fluidity
  – proteins found interspersed among PL bilayer

   integral (intrinsic) proteins - amphipathic, span CM
   peripheral (extrinsic) proteins - located at interior or exterior surface;
      – loosely attached to CM proper

Fluid-Mosaic Model - PL "sea", protein "icebergs"

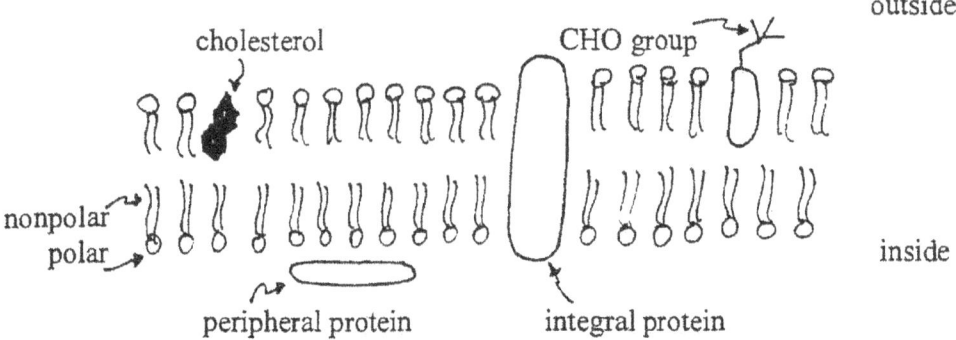

PL arranged in bilayer with polar heads oriented towards aqueous environment found inside/outside cell
glycoprotein - always located at exterior cell surface - used for cell recognition

membrane proteins serve several functions:
  – transport proteins - channel proteins, carrier proteins, enzymes
  – recognition proteins - identify a cell, distinguish self/nonself
  – receptor proteins - substances bind to it, control cellular activity

CM is **selectively permeable (semipermeable)** - regulates molecular traffic

- nonpolar (lipid soluble) molecules - dissolve in lipid bilayer and pass through it
- polar (water soluble) molecules - passage slowed, some require assistance of transport proteins, some excluded by size
- vesicles - some materials are transported across CM in vesicles (endocytosis, exocytosis)

Cellular Organelles -

nucleus - bound by **nuclear membrane** (double membrane with large pores)
- control center of cell, contains genetic material
- **chromatin** - thin threads of DNA + proteins, condense to form **chromosomes** during mitosis

nucleolus - contains rRNA + protein
- assembly site of ribosomal subunits

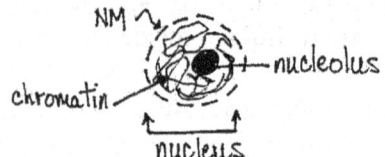

cytoplasm - cell material located between NM → CM
  (½ cytosol + ½ organelles)

cytosol - consists of fluid portion, cytoskeleton and inclusions

cytoskeleton - provides structural framework for cells

   microtubules - hollow cylinders of tubulin protein
   actin filaments - rodlike fibrils of protein
   intermediate filaments - protein fibers

ribosomes - made up of rRNA + protein
- consist of large and small subunits
- attached to ER or free in cytoplasm
- site of protein synthesis

RER - rough endoplasmic reticulum
- membrane system with attached ribosomes
- protein synthesis/storage/transport
- **cisternae** - internal space

SER - smooth ER, no ribosomes present
- manufactures lipids and CHO; detoxification; Ca^{2+} storage

transport vesicle - sac with materials to be conveyed between ER and Golgi

Golgi apparatus - stacked, flattened membrane sacs
- modification, packaging of proteins & lipids for secretion

secretory vesicle - sac pinched off of Golgi apparatus

lysosome - vesicle with digestive enzymes

peroxisome - vesicle with catalase activity
$$2\ H_2O_2 \xrightarrow{catalase} 2\ H_2O + O_2$$

mitochondria - double-membrane organelles with folded inner
　　　　　　　membrane (cristae); site of ATP synthesis

centrioles - cylinders at right-angles, contain MT
　　　　　　– migrate/separate during mitosis

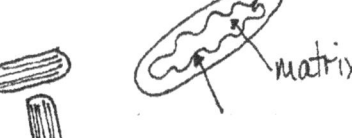

spindle fibers - MT extending from centrioles
　　　　　　– assist chromosome movement/separation

cilia - hairlike, short extensions of CM, contain MT
　　　　　– directed, surface movement

flagellum - long, single extension of CM
　　　　　　– movement; tail of spermatozoa

microvilli - extensions of CM, contain microfilaments
　　　　　　– increase surface area

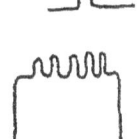

Membrane Transport

Four ways to move material across the CM:
1. Directly through lipid bilayer (materials must be lipid soluble).
2. Through membrane channels (protein pores).
3. Using carrier molecules embedded within the CM (proteins).
4. Within vessicles (endocytosis & exocytosis).

passive processes - no energy requirement
　　　　　　– "downhill" movement of material, obeys concentration gradient
　　　　　　– movement from high → low concentration areas
　　　　　　– e.g., simple diffusion, osmosis, filtration, facilitated diffusion

active transport - requires energy
　　　　　　– "uphill" movement of material, against concentration gradient
　　　　　　– movement from low → high concentration areas
　　　　　　– e.g., active transport (ion pumps, cotransport), endocytosis, exocytosis

Simple diffusion - movements of materials down their concentration gradient
　　　　　　– rate of diffusion increases with increasing concentration gradient & T
　　　　　　　and decreases with increasing size & viscosity
　　　　　　– used primarily to move nonpolar materials across CM
　　　　　　– O_2/CO_2 exchange in the lungs

Osmosis - movement of water across a semipermeable membrane
    – driven by differences in osmotic pressure
    – osmotic pressure = force required to prevent water movement
    – water moves down its concentration gradient and dilutes solute

    isotonic solution = $[H_2O]_{inside} = [H_2O]_{outside}$
        - no H_2O gradient ∴ no net H_2O movement
        e.g., 0.9% NaCl or 5% glucose (isotonic to body fluids)

    hypotonic solution = $[H_2O]_{inside} < [H_2O]_{outside}$ (dilute solution, ↓ solutes)
        - net movement of H_2O into the cell ⇒ **lysis**

    hypertonic solution = $[H_2O]_{inside} > [H_2O]_{outside}$ (concentrated solution, ↑ solutes)
        – net movement of H_2O out of the cell ⇒ **crenation**

Filtration - H_2O and dissolved materials move across the CM due to hydrostatic P
    – eg., capillary dynamics

Dialysis - doesn't occur naturally within the body; used in artificial kidney machines

Facilitated diffusion - mediated transport
    – CM carrier proteins assist polar molecule movement
    – specificity
    permease - aids movement of glucose and amino acids down their
        concentration gradient

Na^+/K^+ ATPase pump - movement of Na^+/K^+ across the CM against
    the concentration gradient and with the expenditure of ATP

    H_2O ↑↓ Na^+ ↓ ↑ K^+
    (diffusion gradient)

Co-transport - secondary active transport
    – Na^+ pumped out of cell, and as it diffuses back into the cell, it assists the
    transport of a polar solute (eg., glucose) against its concentration gradient

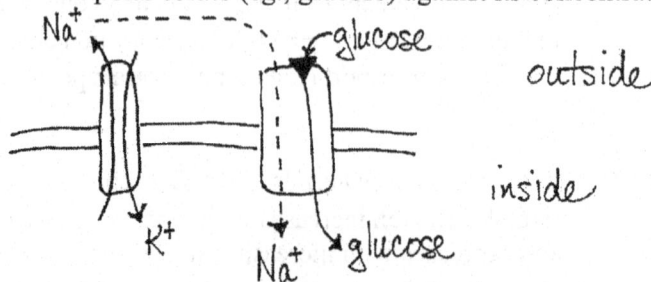

Endocytosis - bulk movement of material into cell with vesicle formation (active)

    phagocytosis - "cell-eating" - large, solid material enters cell
    pinocytosis - "cell-drinking" - large macromolecules enter cell

Exocytosis - secretion of materials from the cell (active)

Cell Metabolism

    aerobic metabolism -   organics $+ O_2 \rightarrow CO_2 + H_2O + \uparrow$ energy

    anaerobic metabolism -   organics $\rightarrow$ lactic acid $+ \downarrow$ energy

Catabolism -

    glycolysis -   glucose $\rightarrow$ 2 pyruvate + 2 ATP
                  occurs within cytoplasm with or without O_2 present

    Krebs cycle -   pyruvate $\rightarrow$ acetyl CoA $\rightarrow CO_2$ + NADH + $FADH_2$
                   occurs within matrix of mitochondria, requires O_2

    oxidative phosphorylation -   NADH/$FADH_2$ are **e^- carriers** that are subjected to a
        (ETS)   series of redox reactions to produce $\uparrow$ ATP
                          – occurs along cristae of mitochondria
                          – requires O_2 as terminal e^- acceptor

Anabolism -

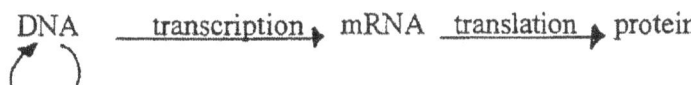

DNA   transcription   mRNA   translation   protein

replication
    (occurs within nucleus)   (occurs within cytoplasm)

mRNA contains 3 base units (codons) that are recognized by complementary 3 base units (anticodons) on tRNA

∴ correct aa carried by tRNA to ribosome and is added to growing peptide chain

Cell Cycle -

    interphase - cell actively metabolizing but not dividing; longest stage of cell cycle
    mitosis - nuclear division in somatic cells; produces 2 identical daughter cells, "clones"

    diploid - 2N = 23 pairs of chromosomes = 46 chromosomes
    haploid - 1N = 23 chromosomes

interphase - time between nuclear divisions; G_1, S & G_2 phases
    G = gap period, interval of growth
    S = synthesis, period of DNA replication

Mitosis -
    prophase - chromatin → chromosomes (2 chromatids joined by centromere)
        – centrioles move to opposite poles; spindle grows between centrioles
        – nucleolus disappears; NM disintegrates
        – longest stage of mitosis

    metaphase - chromosomes align at equator

    anaphase - chromatids separate and move poleward; movement directed by spindle

    telophase - reversal of prophase events; chromosomes → chromatin
        – NM/nucleolus reappear; spindle disappears

    cytokinesis - cytoplasmic division
        – begins late in anaphase with formation of cleavage furrow

Meiosis - produces germ cells (1N); involves 2 consecutive nuclear divisions

    meiosis I - reduction division   results in 4 cells with haploid
    meiosis II ≈ mitosis   chromosome number

    synapsis - occurs during meiosis I, crossing-over permits genetic variability

Cellular aging – cellular clock, death genes, DNA damage, free radicals, mitochondrial damage

Chapter 4 - Histology

Epithelial Tissue - little extracellular matrix
- covers/lines body surfaces
- free border
- basement membrane attaches it to underlying tissue
- avascular - gases/nutrients must diffuse across basement membrane

Classification - based upon # of cell layers & cell shape
- **simple** - monolayer
- **stratified** - multilayer
- **pseudostratified** - false appearance of being multilayered; each cell is in contact with the basement membrane

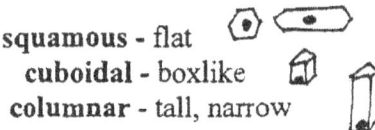

- **squamous** - flat
- **cuboidal** - boxlike
- **columnar** - tall, narrow

stratified squamous - may be **moist** (mouth, esophagus, vagina, rectum) or
keratinized - outer cells dead & contain tough, moisture-resistant protein = **keratin**

tansitional epithelium - stratified, stretches without tearing
- cell layers slide past one another (urinary bladder)

Functions: associated with the # layers & the shape of cells (Table 4.2)

- **simple** - diffusion (lungs), secretion (glands), filtration (glomerulus), absorption (intestines)
- **stratified** - protection/barrier function
    - found in areas where abrasion occurs
- **squamous** - thin/flat → secretion or absorption
- **cuboidal** &/or **columnar** - secretion or absorption

goblet cells - produce mucus
microvilli - increase surface area (brush border of small intestine)
cilia - directed movement across cell membrane (trachea, nasal cavity)

Cell connections - mechanically bind cells together
- form a permeability barrier
- permit intercellular communication
- epithelial cells produce glycoprotein that connects cells to each other and to the basement membrane

desmosomes - disk-shaped structures with sticky glycoproteins
- reinforce cell-cell connections at stress points

hemidesmosomes - attach epithelial cells to the basement membrane

tight junctions - form a permeability barrier
- material can't pass between cells **but** must pass through the cells instead

a) **zona adherens** - girdle of glycoproteins, binds cells together
b) **zona occludens** - ring around cells

gap junctions - small protein channel, allows small polar materials to pass between cells
- important coordinating function (e.g., intercalated disks - cardiac muscle)

Glands - secretory function, composed primarily of epithelia

exocrine - have ducts   **endocrine** - ductless, produce hormones

unicellular - goblet cell → mucus
multicellular - classification based on duct structure & secretion
  simple - duct with few branches
  compound - duct with many branches
  tubular - duct ends in straight/coiled tubes
  acinar - duct ends in cluster of small sacs
  alveolar - duct ends in hollow sacs

  merocrine - secretion with no loss of cytoplasm (e.g., eccrine sweat glands)
  apocrine - secretion with some loss of cytoplasm (e.g., mammary glands)
  holocrine - cell becomes part of the secretion (e.g., sebaceous glands)

Connective Tissue - abundant nonliving, extracellular **matrix**
- protein fibers present ⎫
- ground substance present ⎬ matrix components
- fluid present ⎭
- generally, well-vascularized

- **blasts** - cells that build/create matrix
- **cytes** - cells that maintain matrix
- **clasts** - cells that breakdown/remodel the matrix

Fibers - **collagen** - strong & flexible, doesn't stretch, primary protein in the body
  reticular - fine collagen fiber network
    elastin - stretch & recoil (return to original shape)

Ground Substance - hyaluronic acid - polysaccharide, slippery, acts as a lubricant
proteoglycan - CHO + protein, traps H_2O → resiliency

Fluid - all extracellular fluid associated with a tissue

CT Classification - based on nature of matrix & fibers present (Table 4.3)

1) **Fibrous CT** - loose vs. dense; based on [protein fibers]
   - regular vs. irregular; based on fiber arrangement

   loose CT = CT proper = areolar
   - protein fibers form lacy network with fluid-filled spaces, all three types of fibers present
   - loose packing material of most organs, "space-filler"
   - most abundant CT in body
   - **fibroblast** - cell that produces fibrous matrix of CT

   dense CT - fibers fill nearly all of the extracellular space
   regular dense CT - fibers run in a single direction
   collagenous tissue - seen in tendons & ligaments
   elastic tissue - seen in ligaments of vertebrae

   irregular dense CT - fibers run in many directions
   collagenous irregular - reticular layer of dermis
   elastic irregular - walls of large arteries

2) **Adipose** tissue - fat cells (**adipocytes**) surround lipid droplet
   - energy storage, insulation, protection

3) **Reticular** tissue - network of fine collagen fibers forms framework of lymphoid tissue, bone marrow & liver

4) **Bone marrow** - **yellow** - fat storage
   red - blood cell production (hemopoiesis)

5) **Cartilage** - rigid matrix, lacunae surround chrondrocytes (**cartilage cells**), avascular

   hyaline - rigid with some flexibility, principle cartilage of body
   e.g., costal cartilage, articular cartilage, fetal skeleton,
   trachea, epiphyseal plate, nasal septum, etc.

   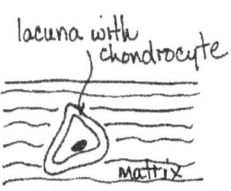

   fibrocartilage - withstands great pressure (shock absorber)
   e.g., intervertebral discs, meniscus of knee, pubic symphysis

   elastic - ↑[elastin], most flexible; e.g., pinna, epiglottis

6) **Bone** - solid matrix with organic & mineralized inorganic material

   inorganic matrix - hydroxyapatite crystals contain Ca/P - → rigidity

   organic matrix - primarily collagen → flexible strength

   cancellous (spongy) bone - bone spicules arranged in trabecular network

   compact bone - dense bone with regular repeating structural units = **osteon**
   (Haversian system)
   - canaliculi
   - lacuna with osteocyte
   - Haversian canal
   - lamellae

7) **Blood** - liquid matrix = plasma

    RBC (erythrocyte) - O_2/CO_2 transport

    WBC (leukocyte) - defense

    platelet (thrombocytes) - blood clotting

Muscle Tissue - excitable tissue and contracts

   Skeletal - striated, voluntary, multinucleate, cylindrical fibers, attached to bones, responsible for body movement

   Cardiac - striated, involuntary, branched fibers, 1-2 nuclei/cell, intercalated discs

   Smooth - nonstriated, involuntary, walls of hollow organs, spindle-shaped fibers

Nervous Tissue - irritable, conducts electrical impulses (action potentials)

   neurons - conducting cells

   neuroglia - supportive, protective cell, insulate neurons

   <u>axon</u> - carries signal away from soma
   <u>dendrites</u> - carry signal towards soma
   dendrites   axon
   cell body (soma)

Embryonic Tissue Development - 3 primary germ layers

    endoderm - inner layer → some epithelia
    mesoderm - middle layer → muscle & CT
    ectoderm - outer → nervous tissue & epidermis

Membranes - cover a structure or line a cavity; have an epithelia & a CT component

    serous membranes - line cavities that do not open to outside; produce serous fluid

    mucous membranes - line cavities that open to outside; secrete mucus

    synovial membranes - line joint cavities; secretes synovial fluid

Tissue Repair - substitute new live cells for existing dead cells

    regeneration - new cells are of same type; function restored

    replacement - new tissue is nonfunctional CT that serves as a space filler →
        causes scar, loss of function

Ability to regenerate:

labile - divide mitotically throughout life (e.g., skin) ∴ can regenerate

stable - cells replicate until growth stops; will regenerate if injured (e.g., CT, glands)

permanent - cannot replicate, replaced by nonfunctional CT (e.g., muscle & neurons)

Tissue damage stimulates the **inflammatory response** - isolates injurious agents;
        attacks/destroys agents
Symptoms include:
- redness -
- heat -
- swelling (edema) -
- pain -
- loss of function -

edema - increase in tissue fluid due to movement of blood proteins into damaged
    tissue

Stages in wound repair:

clot → scab → granulation tissue appears → scab falls off, granulation tissue replaced by
        permanent tissue

Chapter 5 - The Integumentary System

Functions: protection - from microorganisms & certain chemicals
- prevents water loss
- protects against abrasion & UV damage

T regulation - evaporation of sweat produced by sweat glands = evaporative cooling
- vasoconstriction (heat conserved) & vasodilation (radiative cooling) of superficial blood vessels

sensation - sensory receptors embedded within skin layers

excretion - sweating removes small amounts of waste products

vitamin D production - with UV exposure, cholesterol derivative → vit D precursor → liver & kidneys for completion

Skin = **epidermis** + **dermis**   (Table 5.1)

Hypodermis = subcutaneous tissue = superficial fascia
- not part of the skin proper
- attaches skin to underlying bone & muscle
- supplies skin with blood vessels & nerves
- contains loose CT + adipose
- major cells include: fibroblasts, adipocytes, macrophages
- ~ ½ body fat stored within hypodermis

Epidermis - 4-5 layers of stratified squamous epithelium
- primary protective function
- avascular, nourished via diffusion from dermis below
- separated from dermis by basement membrane
- includes accessory structures - hair, nails, oil/sweat glands

Stratum corneum - outermost layer, tough dead cells containing keratin, 25+ layers thick
- undergoes desquamation - cells sloughed off continually

Stratum lucidum - dead, transparent cells, found only in "thick skin", palms & soles

Stratum granulosum - granular appearance, cells filled with keratohyalin
- cell death occurs within this layer

Stratum spinosum - cells held together by many desmosomes

Stratum basale - growing/germinal layer
- cells undergo mitosis & push toward surface (takes ~ 40-60 days)
- connected to underlying dermis via basement membrane

thick skin = all 5 layers, no hair; thin skin = 4 layers, no stratum lucidum, hair present

stratum basale contains: **keratinocytes** - produce waterproofing protein = keratin;
melanocytes - produce pigment **melanin**, transferred to keratinocytes via exocytosis;
melanin - absorbs UV light, protects dividing cells against UV damage

keratinization - living stratum basale cells → dead, keratin-filled stratum corneum cells

keratinized cells - keratin-filled with protein envelope; held together by desmosomes; contributes structural strength to the skin; intercellular space filled with lipids → waterproofing quality to skin

soft keratin = within skin & inside hairs; hard keratin - nails & outside of hairs

Skin color - determined by: skin pigments; blood circulation; thickness of st. corneum

- Melanin - brown/black pigment, produced by melanocytes
  - packed into melanosome vesicles, released by exocytosis
  - production determined by genetics, hormones, UV exposure
  - albinism - genetic lack of melanin production
  - racial variation in skin color due to [melanin] produced, # melanocytes ~ same in all races

- Blood flow - dilation of superficial blood vessels increases blood flow to skin → red/pink skin color
  - cyanosis - skin appears blue due to ↓ O_2 content of blood

- Cartenoids - contribute yellow cast to the skin

Langerhans cells - found in epidermis; immune function

Dermis - inner layer of skin
  - provides most of structural strength to the skin
  - primarily dense, irregular CT with fibroblasts, macrophages & adipocytes

Papillary layer - contains rich blood & nerve supply; nourishes epidermis
  papillae - projections of dermal ridges into epidermis above
    - adaptation for fine grasping; give rise to fingerprints

Reticular layer - main fibrous layer, ↑ fiber content → flexible strength;
  - continuous with the hypodermis below
  - cleavage or tension lines - due to orientation of collagen/elastin fibers
  - with surgery - incision across these lines → gaps, scar formation
    - incision parallel to these lines → less gap/scar formation
  - stretch marks - if skin is overstretched, dermis ruptures & lines are visible through the epidermis

transdermal patches - drug delivery through the skin
- drugs diffuse through epidermis → dermal blood vessels
- won't work for every drug - solubility problems

Burns - rule of 9s - used to estimate % body surface affected
  1st degree - epidermis only, red/painful, slight edema - sunburn
  2nd degree - epidermis/dermis damaged, 1st degree symptoms + blistering
       - superficial vs. deep, may scar
  3rd degree - skin destroyed, scarring, life-threatening, skin grafts

Hair - shaft - above skin surface, dead keratinized cells, arise from hair follicle
     root - part of hair below skin surface
     bulb - @ base of root; epithelial cells undergo division

  - experiences growth stage (~3 yrs) followed by resting stage (1-2 yrs)
  - after growth/rest cycle hair falls out & process begins again
  - normal hair loss ~ 100 hairs/day
  - hair color determined by [melanin]; red hair - melanin with Fe
  - gray/white hair - ↓ [melanin]

Arrector pili - smooth muscle associated with hair follicle
          - contracts to pull follicle downward, "hair stands on end" → gooseflesh

Nails - stratum corneum with hard keratin, protects ends of digits, used in grasping

Glands -
  Sweat (sudoriferous) glands - function in excretion & thermoregulation
    - merocrine (eccrine) - used in thermoregulation, empty directly onto epidermis
    - apocrine - empty into hair follicles of axillary & genital region
        - secretion contains organic material as well as water

  Oil (sebaceous) glands - release sebum (softens & waterproofs)
    - an example of a holocrine gland
    - associated with hair follicles

Chapter 6 - The Skeletal System

Skeletal system - capable of growth, adapts to stress, repairs after injury
- includes bone, cartilage, tendons & ligaments

Functions of skeletal system - support
- protection - skull, vertebrae, rib cage
- allows movement
- storage - minerals (Ca & P), fat (within yellow marrow)
- blood cell production (hematopoeisis) - bloods cells & platelets produced in red bone marrow

Tendons - band or cord of dense regular CT, connects muscle → bone

Ligaments - dense regular CT, connects bone → bone

tendons & ligaments - contain bundles of collagen fibers, few nerves/blood vessels

grow by
- **appositional growth** - new matrix/cells added to outside of tissue
- **interstitial growth** - cells within tissue divide and add more matrix from the inside

Hyaline cartilage - chondroblasts → chondrocytes (within lacuna)
- matrix contains: **collagen** - provides flexible strength; **proteoglycan** - traps water, provides *resiliency* (returns to original shape after compression)
- found at articular surfaces
- replaced by bone in fetal skeleton
- avascular - nutrients reach chondrocytes by diffusing through matrix
- grows by appositional & interstitial growth

Bone - classified according to shape: **long** - femur
　　　　　　　　　　　　　　　　short - carpals/tarsals
　　　　　　　　　　　　　　　　flat - skull, ribs, sternum
　　　　　　　　　　　　　　　　irregular - vertebrae, sphenoid, facial bones

Bone matrix - 35% organic - collagen for flexible strength & proteoglycans
　　　　　　　65% inorganic - hydroxyapatite (3 $Ca_3(PO_4)_2 \cdot Ca(OH)_2$)
　　　　　　　　- hardness, weight-bearing strength

diaphysis - shaft of long bone
epiphysis - ends of long bones
periosteum - double layer of dense fibrous CT, covers outer bone surface
- inner layer has osteoblasts, site of growth in bone diameter
endosteum - lining of marrow cavity, contains osteoblasts
articular cartilage - hyaline cartilage covering bone ends @ joints
epiphyseal plate - hyaline cartilage between epiphysis & diaphysis
- site of vertical bone growth, endochondral ossification
cancellous (spongy) bone - found within epiphyses, arranged in trabeculae
compact (cortical) bone - dense bone arranged into osteons
- found throughout diaphysis & covering spongy bones of epiphyses
medullary cavity - cavity within diaphysis
red marrow - found within spaces of cancellous bone, site of hematopoeisis
yellow marrow - fat stored within medullary cavity

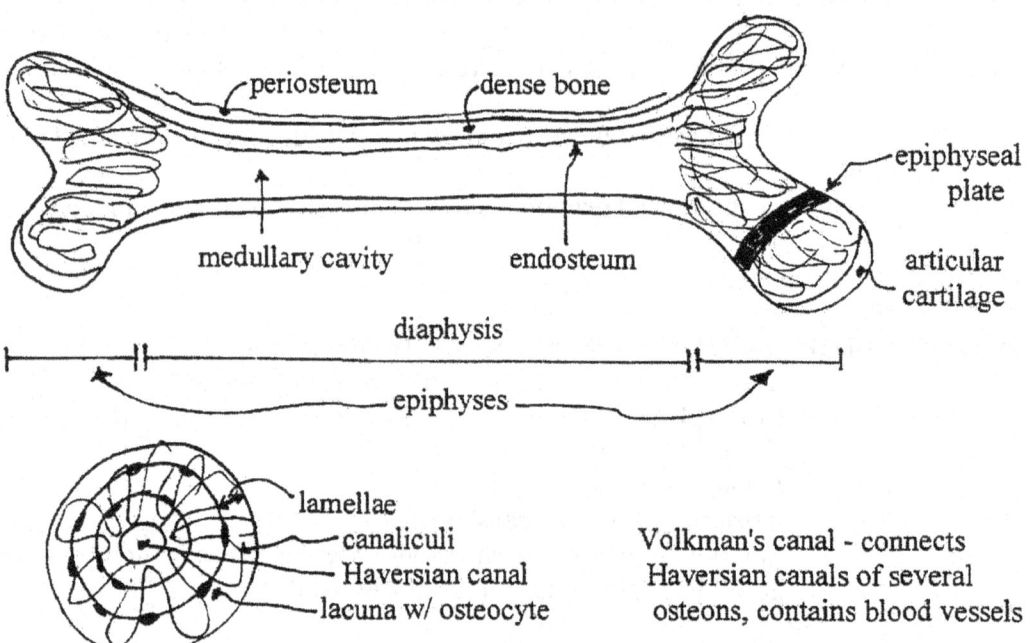

ossification - process of bone formation
- osteoblasts make organic matrix & minerals are added
- intramembranous & endochondral processes form both spongy and compact bone

woven bone - immature bone, collagen fibers oriented in different directions
- replaced by mature bone with collagen fibers organized to form lamellae (remodeling)

Two mechanisms of fetal bone development -

intramembranous ossification - results in flat bones
- occurs within membranes, osteoblasts produce cancellous bone
- beneath periosteum, osteoblasts lay down compact bone to form outer shell
- **fontanels** - "soft spots" - regions of incomplete intramembranous ossification in fetal/newborn skull

endochondral ossification - produces long bones
- bones develop from within a cartilage model/template
- cartilage is then calcified and dies
- osteoblasts form bone on calcified cartilage → spongy bone
- outer surface of compact bone is formed beneath periosteum
- 1° ossification centers form in diaphysis during fetal development
- 2° ossification centers form in epiphyses → epiphyseal plate
- epiphyseal plate closes completely between 18-21 years of age, no more vertical possible
- articular cartilage remains at ends of bones

Bone growth - bone grows by either appositional or endochondral processes (can **not** grow via interstitial method)

appositional growth - increases diameter of long bones & is responsible for most growth of other bones
- osteoblasts on surface of bone divide → surrounded by matrix → trapped within lacuna (osteocytes)

endochondral growth - increases length of long bones at epiphyseal plates
- interstitial growth of hyaline cartilage within plate occurs
- with calcification, cartilage cells die and are replaced by bone → endochondral ossification

zones of epiphyseal plate - zone of resting cartilage   epiphysis
zone of proliferation
zone of hypertrophy
zone of calcification   diaphysis

factors affecting bone growth: vitamin D - important for Ca^{2+} absorption
vitamin C - important for collagen synthesis
hormones - GH, T_3/T_4, sex hormones

Ca^{2+} homeostasis - 2 hormones help regulate $[Ca^{2+}]$ in blood

    PTH (parathyroid hormone) - stimulates osteoclast activity → bone resorption
- increases Ca^{2+} absorption in small intestines
- increases Ca^{2+} reabsorption in kidneys

∴ PTH → ↑ blood $[Ca^{2+}]$

    calcitonin - stabilizes osteoclasts, promotes osteoblast activity
∴ calcitonin → ↓ blood $[Ca^{2+}]$

Bone remodeling - involves activity of osteoblasts & osteoclasts
- when stress is applied to bone, growth occurs on same side as stress & resorption occurs on opposite side (e.g., braces to straighten bones)
- inactivity causes bone resorption - jawbones recedes, etc.
- physical exercise & mechanical stress promote bone growth

Bone repair - fracture repair begins with blood clot (hematoma) within first few days
- fibrocartilaginous callus forms a zone of repair, within 1 week
- osteoblasts (from periosteum/endosteum) enter callus and convert it to osseus callus, within 4-6 weeks
- bone is remodeled

Fracture types - simple - damage localized within skeletal system
- compound - bone pierces through skin
- complete - bone is broken in two pieces
- green stick - bone fragments still attached
- linear (↓), transverse (→), oblique (↗), spiral (ϑ)
- comminuted - may pieces
- compacted - one bone piece driven into another
- compression - due to excessive vertical force, e.g., intervertebral disks

Bone disorders -
    rickets - due to ↓ vitamin D in children; results in soft, deformed bones
    osteomalacia - "softening" of adult bones due to ↓ [Ca/P]
    scurvey - due to ↓ vitamin C; body can't synthesis enough collagen for organic matrix
    arthritis - rheumatoid - autoimmune, attacks small joints (fingers), bilateral effects
        - **osteoarthritis** - with aging, articular cartilage lost, bone spurs develop. usually attacks larger joints (hip, knee)
    osteoporosis - demineralization of bone → brittle bones
- seen primarily in postmenopausal women, treated with Ca^{2+} supplements, vitamins C & D, exercise, hormone therapy

    GH abnormalities - age-dependent disorders
        gigantism - ↑ GH in children   **acromegaly** - ↑ GH in adults
        pituitary dwarf - ↓ GH in children
    osteomyelitis - bone infection/inflammation

Chapter 7 - The Skeletal System: Important "Facts" to Know & Apply

Axial skeleton - skull [cranium (8), facial bones (14), auditory ossicles (6)] + hyoid +
        vertebral column (7C, 12T, 5L, 5S = 1 sacrum, 1 coccyx) +
        thorax [ribs (24) + sternum]

Appendicular skeleton - pectoral girdle (scapula + clavicle) + humerus, radius, ulna,
        carpals, metacarpals, phalanges
        - pelvic girdle (os coxa = ilium, ischium, pubis) + femur, tibia,
        fibula, patella, tarsals, metatarsals, phalanges

Basic features of bone -

Bones of the orbit of the eye - frontal, maxilla, zygomatic, lacrimal, ethmoid, sphenoid,
        palatine

Bones with paranasal sinuses - frontal, ethmoid, sphenoid, maxilla

Bones of the nasal septum - vomer, perpendicular plate of the ethmoid

Bones of the hard palate - palatine process of maxilla (front 2/3) +
        horizontal plate of palatine (back 1/3)

Vertebral column - curved to support more weight
    1° curves - thoracic + sacrococcygeal
    2° curves - cervical + lumbar

Intervertebral disks - annulus fibrosus (external) + nucleus pulposus (internal)
    herniated/ruptured disk - breakage of annulus fibrosus with release inner materia
        - may push against spinal cord/nerves → pain/loss of fcn
        - most common in inferior cervical ("whiplash") &
        inferior lumbar regions

Abnormal spinal curves: lordosis - convex curve in lumbar area (swayback)
        kyphosis - concave curve in thorax (humpback)
        scoliosis - abnormal bending of spine to the side ("S" or "C")

Sacroiliac joint - receives most of the weight of the upper body
        - forward weight (eg. pregnancy) stretches CT & nerve endings → pain

Differences between male & female pelvis:

♂ { larger, heavier / funnel-shaped (tall, narrow) / pelvic inlet = heart-shaped / subpubic △ less than 90° }

♀ { smaller, lighter / basin-shaped (broader) / pelvic inlet = oval-shaped / subpubic △ more than 90° }

Chapter 8 - Articulations

3 Types of Articulations:
1) <u>fibrous joints</u> - bones joined by fibrous CT
    a) <u>sutures</u> - found between skull bones, immovable in adults
        <u>fontanels</u> - "soft spots" in newborn, allow for growth of head & "give" during birth process
        <u>synostosis</u> - two bones grow together across a joint to form a single bone

    b) <u>syndesmoses</u> - bones joined by interosseous ligaments; slight movement allowed
        - eg: radius-ulna; tibia-fibula distal connections

    c) <u>gomphosis</u> - peg-and-socket joint; teeth fitting into mandible/maxilla

2) <u>cartilaginous joints</u> - bones joined by hyaline or fibro-cartilage
    a) <u>synchondroses</u> - bones joined by hyaline cartilage
        <u>epiphyseal plate</u> - temporary synchondrosis; no movement allowed
        <u>costosternal synchondroses</u> - between ribs & sternum using costal cartilage; must "give" to allow for respiratory movements

    b) <u>symphysis joint</u> - fibrocartilage present; found in pubic symphysis & intervertebral discs; allow for slight movement

3) <u>synovial joints</u> - enclosed by a joint capsule consisting of outer fibrosus capsule + inner synovial membrane
    <u>synovial membrane</u> - produces synovial fluid, slippery consistency, lubricating fluid, supplies nutrients to joint components
        <u>articular cartilage</u> - hyaline cartilage
        <u>articular discs</u> - fibrocartilage; meniscus of knee
        <u>bursa</u> - extension of synovial membrane to form a pocket

    - synovial joints are classified according to movement and structural types:

    <u>movement types</u>: a) monoaxial - movement in 1 direction/plane
                b) biaxial - movement in 2 directions/plans
                c) multiaxial - movement in all 3 directions/planes

structural types:
- a) plane/gliding - 2 flat surfaces, monoaxial with some rotation possible; articulation between vertebrae; between carpals
- b) saddle - biaxial; carpometacarpal joint of thumb; permits opposition
- c) pivot - monoaxial; rotation around single axis; dens of axis & atlas; proximal ends of radius & ulna
- d) hinge - monoaxial; humerus & ulna; knee; interphalangeal joints
- e) ball-and-socket - multiaxial; hip & shoulder
- f) ellipsoidal - biaxial; atlantooccipital; radiocarpal; temporomandibular joints

Range of motion:
1- flexion - reduce ∠ between bones; extension - increase ∠ between bones hyperextension - extension beyond normal range of motion
2- dorsiflexion - stand on heels; plantar flexion - stand on toes; flex & point
3- abduction - away from midline; adduction - toward midline
4- rotation - turn structure around its long axis; side-to-side head movement
5- pronation - palm faces dorsally; supination - palm up, faces anteriorly
6- circumduction - combination of flexion/extension & abduction/adduction; describe a cone with should or hip
7- elevation - move superiorly; depression - move inferiorly; open/close mouth; shrug shoulders
8- lateral/medial excursion - move mandible to right/left of midline (lateral) and return to neutral position (medial)
9- opposition - touch thumb to digits; reposition - return thumb to neutral position
10- inversion - ankle with sole facing medially; eversion - sole facing laterally
11- protraction – moving a structure in an anterior direction; retraction – move back to anatomical position or even more posteriorly; mandible & scapula

Some special features of joints:
TMJ - normal working requires functional dislocation

shoulder vs. hip - shoulder allows for more range of motion but has less bony protection/stability; shoulder is easier to dislocate than hip

knee - actually more than a simple hinge joint (more complex ellipsoidal); allows flexion/extension with some rotation; contains lateral/medial meniscus, tibial/fibular collateral ligaments; anterior/posterior cruciate ligaments; most knee injuries come from side blows & damage medial components

ankle - talocrural joint; modified hinge joint

arches of foot - distribute weight of body between heel/ball of foot; formed as a result of ligaments and muscles acting on the feet

Chapter 9 - The Muscular System: Histology/Physiology

muscle exhibits: **contractility** - shortens when stimulated
    excitability - responds to stimulus by nerves/hormones
    extensibility - can be stretched beyond their resting length to some degree
    elasticity - has both stretch and recoil capabilities

** review comparison between types of muscle tissue **

muscle cells = muscle **fibers**   1-40 mm in length/ 10 - 100 μm in diameter

myoblasts - produce muscle fibers; # muscle fibers ~ constant after birth

hypertrophy - involves increase in muscle fiber size (more protein) rather than in number

Connective tissue "wrappings" associated with muscle:
  external lamina - reticular fiber layer surrounding muscle fibers
  sarcolemma - CM of muscle fiber
  endomysium - loose CT surrounding muscle fibers (outside external lamina)
  perimysium - surrounds a bundle (fascicle) of muscle fibers
  epimysium - many fasicles grouped together
  fascia - fibrous CT outside of epimysium

muscle fibers contain: sarcolemma, sarcoplasm, sarcoplasmic reticulum (SER)

myofibrils - threadlike structures composed of 2 **myofilaments**
    myosin - thick filaments   12 nm wide x 1800 nm long
    actin - thin filaments   8 nm wide x 1000 nm long

sarcomere - functional unit of muscle

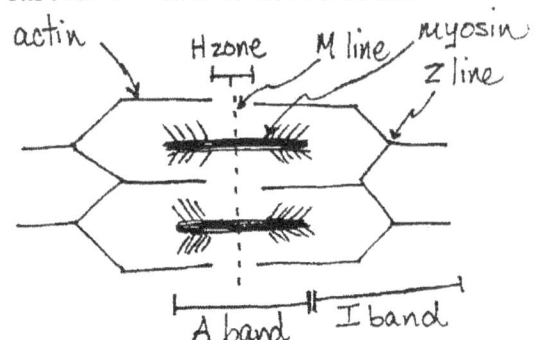

I = isotropic ≡ light

A = anisotropic ≡ dark

Each **actin** myofilament contains - 2 fibrous actin (F-actin) + tropomyosin + troponin
    actin strands coil to form a double helix
    each F-actin polymer ≈ 200 globular actin (G-actin) monomers
    each G-actin monomer has a binding site for myosin

tropomyosin - regulatory protein, "covers" 7 G actin binding sites
troponin - has 3 binding sites: 1 for actin, 1 for tropomyosin, 1 for Ca^{2+}

myosin - rod-shaped molecule with 2 heads (contain ATPase)

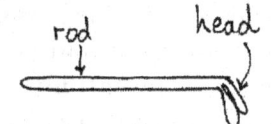

cross bridge - myosin head + active site of sctin

T-tubules (transverse) - tubelike invaginations of sarcolemma

sarcoplasmic reticulum - smooth ER with enlarged terminal cisternae for Ca^{2+} storage
- releases Ca^{2+} when muscle stimulated to contract

Sliding filament theory: actin & myosin **don't** change length during contraction
- they **slide past one another**
- x-bridges form between actin/myosin → x-bridges **form, move, release, reform**
- action results in inward movement of actin filaments towards H zone
- Z lines are brought closer together

Contraction = active process, ATP required
Relaxation = passive process, x-bridge release, no ATP required

Physiology of Skeletal Muscle Fibers

   motor neurons - nerve cells that propagate action potentials (APs) to skeletal muscle
                 (for discussion of AP, see Chapter 11)
   neuromuscluar junction -
- synaptic vesicles contain NT acetylecholine (Ach)
- AP at presynaptic terminal causes Ca^{2+} channel to open which causes vesicles to migrate
- NT released into synapse
- Ach binds to protein receptor on postsynaptic side
- ligand-sensitive Na^+ channels open
- AP is propagated along sarcolemma

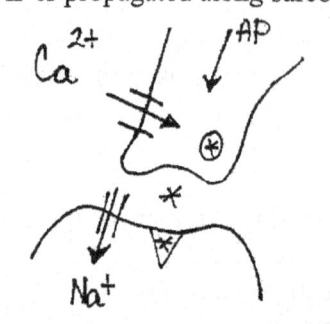

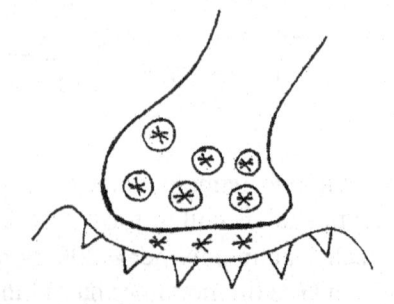

acetylcholinesterase - breaks down unused Ach within synaptic cleft

$$Ach \xrightarrow{Achase} acetic\ acid + choline$$

prevents constant stimulation of the junction; 1 presynaptic AP → 1 postsynaptic AP

Excitation contraction coupling - mechanism by which AP results in muscle contraction

- AP reaches T tubules (carries depolarization to interior of muscle fibers)
- Ca^{2+} diffuses into sarcoplasm surrounding myofibrils
- Ca^{2+} binds to **troponin**
- **troponin-tropomyosin complex** "moves" deeper into groove between 2 F-actin molecules
- active site on actin is exposed
- **actin-myosin x-bridge** form

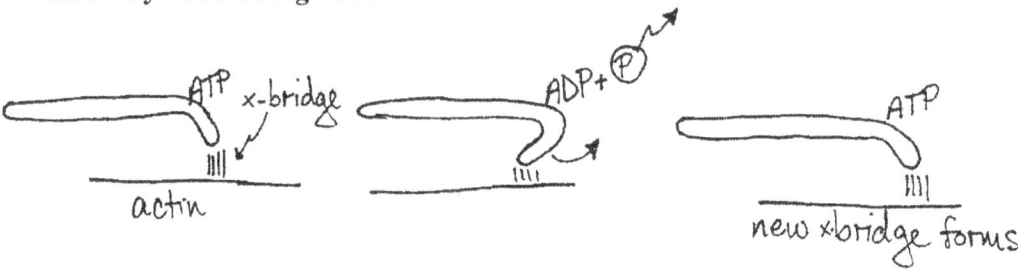

1 ATP required for x-bridge formation/movement/release cycle
ATP stored within myosin head; ATPase breaks down ATP → ADP + P_i

power stroke - movement of myosin head while x-bridge is in place

recovery stroke - return of myosin head to its original position after x-bridge broken

muscle relaxation - Ca^{2+} reabsorbed into sarcoplasm reticulum
- Ca^{2+} leaves troponin
- troponin/tropomyosin move back into resting position
- actin active site is covered up

Muscle twitch - contraction of muscle in response to a stimulus that causes an AP in one or more fibers

lag phase - time between stimulus at motor
   neuron & beginning of contraction

contraction - time of contraction
   phase   (muscle shortening)

relaxation - time of relaxation (recovery)
   phase

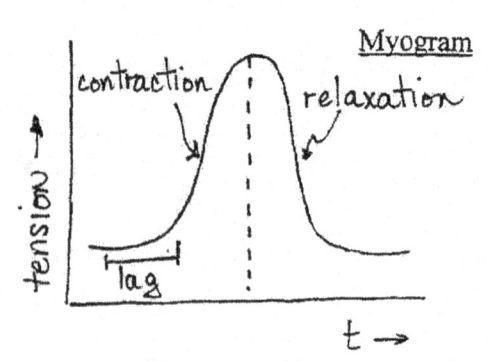

Myogram (diagram: tension vs t, showing lag, contraction, relaxation)

AP = electrochemical event   **contraction = mechanical event**
   (duration = ~1-2 msec)   (duration = ~1 sec)

all-or-none law of contraction - with an appropriate stimulus received, muscle fibers produce contractions of equal force

- **subthreshold stimulus** - no AP produced, no resultant contraction

- **threshold stimulus** - AP produced, muscle contraction results

- **above-threshold stimulus** - AP produced, muscle contraction of the same magnitude as that produced by a threshold stimulus results

motor units = 1 motor neuron and all the muscle fibers it innervates

- **whole** muscles are composed of many motor units

- **axons** of motor units combine to form **nerves**

- whole muscles respond in a **graded** rather than all-or-none fashion

maximal stimulus - all motor units of a muscle are activated

- as stimulus strength is increased from subthreshold → threshold → submaximal → maximal strengths, additional motor units are **recruited; recruitment**

- more motor units responding → ↑ contraction force & tension

- muscles with delicate/precise movement have motor units with ↓ # muscle fibers → results in more control; small ratio; 1 motor neuron:10-12 muscle fibers (cells)

- muscles which generate more powerful movements have motor units with ↑ # muscle fibers/motor unit → results in more muscle power/strength; large ratio;
1 motor neuron:1,000 + muscle fibers (cells)

muscles fibers don't have to completely relax before stimulated by a 2nd AP

    incomplete tetanus - fibers partially relax between contractions

    complete tetanus - APs occur so rapidly that there's no time for muscle relaxation between APs

    multiple-wave summation - increased tension caused by increase in stimulus frequency

    tetanus is caused by stimuli of increasing frequency

    treppe - a "staircase"-like graded response
- a "warming-up" phenomenon in rested muscle
- first few APs result in contraction of increasing strength/tension
- all further contractions show same tension

Types of muscle contraction –

    Isometric - tension increases during contraction **but** length of muscles does not
- contractions exhibited by postural muscles

    Isotonic - tension is ~ constant during contraction **and** length of muscle changes
- contraction associated with joint movements

Muscle tone - constant tension produced by muscles over long periods of time
- keeps back/legs straight, head upright, abdomen flat
- depends on small # of motor units contracting out of phase with each other (i.e., same motor unit not contracting all the time)

Concentric contractions - muscle produces ↑ tension as it shortens

Eccentric contractions - muscle tension maintained while muscle increases in length
(e.g., lowering of a heavy weight slowly)

Active tension - force applied to object to be lifted when muscle contracts
- as length of muscle increases, its active tension also increases
- provides optimal overlap of actin/myosin filaments; x-bridge formation causes maximal contraction

Passive tension - applied to the load when a muscle is stretched but not stimulated
- due to the elasticity of muscle and its connective tissue

Total tension = active + passive tension

Fatigue - decreased capacity to do work
- muscle experiences reduced efficiency
- usually follows period of activity

a) **psychological** - involves CNS, "perceive" additional work is not possible so you consciously shutdown activity (muscle can still function)
- additional burst of activity in athletes due to "crowd" → psychological fatigue can be overcome

b) **muscular** - results from ATP depletion

c) **synaptic** - depletion of synaptic vesicles & NT Ach (rare)

Rigor mortis - develops after death, x-bridges form and can't be broken
- Ca^{2+} leaks out of sacroplasm reticulum, x-bridge forms but inadequate ATP is available for x-bridge release

Physiological contracture - extreme muscle fatigue, incapable of contracting/relaxing, due to ↓ ATP

Energy source for muscle contraction - ATP required for x-bridge dynamics
creatine phosphate - stores energy, used to replenish ATP supply

creatine phosphate + ADP → creatine + ATP

anaerobic respiration - w/o oxygen, lactic acid accumulates
- can be used for short bursts of energy but not for endurance activity

aerobic respiration - w/ oxygen present, glucose broken down to $CO_2 + H_2O$ with the production of large quantities of ATP
- supports endurance activities

oxygen debt - oxygen required after intense activity to reestablish normal concentration of creatine phosphate, ATP and to convert lactic acid → pyruvate (w/in liver)
- pyruvate → glucose → glycogen conversion restores [glycogen] in liver & muscle

slow-twitch (high-oxidative) fibers - "dark meat"
- contract more slowly, smaller fibers, good blood supply, ↑ # mitochondria
- ↑ fatigue-resistant
- aerobic respiration predominant
- contain ↑ [myoglobin]; myoglobin ~ Hb, stores/reservoir for O_2
- myoglobin increases ability of muscle to do aerobic respiration

fast-twitch (low-oxidative) fibers - "white meat"
- respond quickly, less-developed blood supply, ↓ #/size of mitochondria
- fatigue quickly
- ↑ deposits of glycogen
- adapted for anaerobic respiration

sprinters - ↑ # fast twitch fibers   **distance runners** - ↑ # slow twitch fibers

exercise - produces **hypertrophy** of existent muscle fibers
    hypertrophy - increases cell size not cell number
    atrophy - decrease size of muscle with disuse

  intense exercise (anaerobic) = ↑ muscle strength/mass; has a greater effect on
                      fast-twitch fibers
      aerobic exercise = increases # mitochondria, increases blood supply,
                      increases endurance of both slow/fast twitch fibers
            - primarily increases the size of slow-twitch fibers

with training - fast-twitch fibers display more endurance
           - slow-twitch fibers generate more force

heat production - **shivering** - muscle contractions generate heat, increase body T

Smooth muscle - spindle-shaped fibers, mononucleate, involuntary
- **not** organized into sarcomeres, not striated
- slower contraction speed but sustain tension longer
- decreased # actin/myosin filaments present
- contain intermediate filaments & dense bodies
- actin attaches to intermediate filaments (~ equivalent to Z lines)
- Ca^{2+} binds to **calmodulin** and activates myosin kinase
- myosin kinase adds P_i from ATP to myosin (x-bridge formation)
- myosin phosphatase removes P_i from myosin (release of x-bridge)

a) visceral/unitary smooth muscle - found in GI, reproductive and urinary tracts
- occurs in "sheets", equipped with gap junctions (functional synctium)
- some is autorhythmic (GI tract), some must be stimulated (bladder)

b) multiunit smooth muscle - found as sheets (blood vessel walls), in small bundles
    (arrector pili) or as single cells (spleen)
- contractions only when stimulated, few gap junctions

smooth muscle RMP ~ -55-60 mV
- doesn't respond in an all-or-none fashion
- slow wave of depolarization due to Na^+ and Ca^{2+} movement
- epinephrine stimulates smooth muscle contraction by activating G-proteins
- oxytocin stimulates muscle of uterus
- smooth muscle contracts when stretched & exhibits constant tension
- innervated by ANS (involuntary)

Cardiac muscle - branched fibers, striated ~ mononucleate, intercalated disks, autorhythmic, depolarization involves Na^+/Ca^{2+} movement

spastic paralysis - Ach is not degraded and accumulates at synapse
- muscle contracts/can't relax
- e.g., organophosphate pesticides

flaccid paralysis - Ach can't bind to receptor at NM junction ∴ muscle doesn't contract
- e.g., curare, binds to Ach receptors at NM junction

myasthenia gravis - produce Ab to Ach receptors ∴ reduces # of functional receptors
- casues flaccid paralysis
- treated by neostigmine, blocks Achase activity allowing Ach to accumulate in the NM junction & bind to remaining receptors

poliomyelitis - motor unit destruction, causes loss of muscle function in fibers innervated by destroyed nerves

anabolic steroids - increase size/strength of muscles; side-effects include: sterility, cardiovascular problems, liver dysfunction, irritability

muscular dystrophy - destroyed skeletal muscle replaced with CT
  Duchenne - sex-linked inheritance, primarily affects males
  Facioscapulohumoral - less severe form, both males/females affected equally

fibrosis - damaged muscle tissue is replaced by CT
fibrositis - inflammation of fibrous CT → stiffness, pain, soreness
cramps - painful, spastic contractions; caused by ↑ [lactic acid] & fibrositis
fibromyalgia - chronic muscle pain syndrome

Chapter 10 - The Muscular System: Gross Anatomy

Tendons - attach muscle → bone; e.g., cordlike or broad/flat (aponeurosis)

Muscles - **origin** (head) - attached to stationary bone
      insertion - attached to moving bone
      belly - located between origin & isertion

Synergists - muscles that work together to produce a movement
Prime mover - muscle that plays the major role in the observed movement
Antagonist - muscle that works against agonists (prime mover + synergists)
Fixator - stabilizes joint crossed by prime mover

forearm flexion: prime mover = biceps brachii
                synergist = brachialis
                antagonist = triceps brachii

Classification of muscles: location (pectoralis)
                        size (maximus, minimus)
                        shape (deltoid)
                        fiber orientation (oblique, rectus)
                        origin/insertion (sternocleidomastoid)
                        number of heads (biceps, triceps)
                        function (flexor carpi radialis)

Force = pull applied to lever as a result of muscle contraction
Lever = bone, a rigid shaft capable of turning about a pivot point (= joint)
Weight = resistance; object that the force's movement is directed at
Fulcrum = joint, pivot point for observed movement

Class I Lever   <u>weight</u>   <u>fulcrum</u>   <u>force</u>

    atlantoccipital jt (fulcrum), posterior neck muscles (force), head (weight)
    action: depressing back of head; limited distance or amt of weight lifted

Class II Lever   <u>fulcrum</u>   <u>weight</u>   <u>force</u>

    calf muscles (force), calcaneus (lever), weight of entire body (weight),
    metatarsal-phalangeal joint (fulcrum)
    action: standing on your toes; lift considerable weight but only a short distance

Class III Lever   <u>fulcrum</u>   <u>force</u>   <u>weight</u>

    biceps brachii (force), ulna (lever), weight of forearm/hand (weight)
    action: elbow flexion; greater distances of movement, not much weight lifted
    ** most common lever system in body **

Chapter 11 – Functional Organization of Nervous Tissue

CNS = brain + spinal cord

PNS = nerves + ganglia
    nerves - bundle of axons extending from CNS → periphery
        – 12 pair of cranial nerves & 31 pair of spinal nerves
    ganglia - collection of neuron cell bodies located outside CNS

Divisions of the PNS
1- **afferent** (sensory) - sensory organs → CNS
       – cell bodies located in ganglia near spinal cord/cranial nerve origin

2- **efferent** (motor) - CNS → effector (muscle, gland)

Divisions of efferent division of PNS
a) **somatic motor nervous system (SNS)** - CNS → skeletal muscle
       – neuron cell bodies located within CNS; axons extend to NM junction
       - somatic reflexes

b) **autonomic nervous system (ANS)** - CNS → smooth/cardiac muscle & glands
       – provides subconscious control
       – has 2 sets of neurons:
          cell bodies of 1st neuron within CNS → axon extends to autonomic ganglia
          cell bodies of 2nd neuron in autonomic ganglia → send axon to effector
       - visceral reflexes

Divisions of ANS

    sympathetic - prepares body for activity/stress
    parasympathetic - regulates resting/vegetative functions; return to homeostasis

CNS - **processing** information, **initiating responses, integrating** mental processes

PNS - **detect** stimuli, **transport** information to/from CNS in the form of APs

Neurons - receive stimuli/conduct AP

    cell body (soma) - contains nucleus, RER, golgi, mitochondria, neurofilaments
    dendrites - short, branched processes with extensions (dendritic spines)
        – carry information **toward** cell body
    axon - (= nerve fiber) - begin at axon hillock, single process
        – carries information **away from** cell body
    Nissl bodies - found within cell body; specialized areas of RER → protein synthesis

within axon - **axoplasm, axolemma, telodendria** (branches at axon terminus)
with enlarged ends called **presynaptic terminals (terminal boutons)**

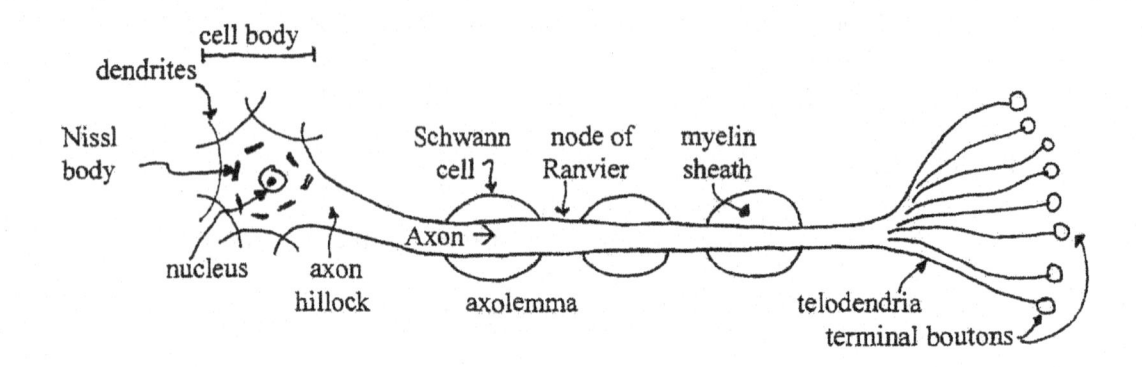

- axon transport moves proteins, organelles (mitochondria) & vesicles down the axon → terminus
- materials can also be transported up axon → cell body (e.g., viruses, chemicals)

Types of Neurons
        Afferent/Sensory   - sensor → CNS
        Efferent/Motor   - CNS → effector
        Association/Interneuron   - between neurons within CNS

Multipolar neurons - 1 axon, many dendrites   e.g., association & motor neurons
Bipolar neurons - 1 axon, 1 dendrite   e.g., sensory neurons of rods/cones
Unipolar neurons - 1 process = axon; peripheral ends have dendrite-like processes
        e.g., most sensory neurons

Neuroglia - supporting cells in CNS; # glia > # neurons; ~ ½ brain's weight

1- **astrocytes** - star-shaped; cover surfaces of blood vessels, neurons, pia mater
    – regulate extracellular fluid around neuron
    – important in **blood-brain barrier** structure/function

    blood-brain barrier - very selective, protects CNS from toxins
      – nonpolar (lipid-soluble) materials pass through barrier, e.g., nicotine, EtOH
      – polar (water-soluble) materials can't pass without help (mediated transport)
        e.g., aa & glucose

Parkinson's disease - lack NT dopamine → ↓ muscle control, ↑ shaking
 – administer drug levodopa, L-dopa which crosses blood-brain barrier instead of dopamine which is polar & therefore cannot pass

2- **ependymal cells** - line ventricles (cavities) of brain & central canal of spinal cord

   choroid plexus - specialized ependymal cells that secrete CSF (cerebrospinal fluid) which circulates through the brain/cord
   – some cells contain cilia to move CSF

3- **microglia** - small cells, mobile & phagocytic
   – respond to inflammation & the presence of disease-producing organisms

4- **oligodendrites** - form sheaths around axons within CNS

5- **Schwann cells** or neurolemmocytes - form myelin sheath around axons within PNS

   satellite cells - specialized neurolemmocytes that surround soma in ganglia
   – give support/provide nutrients

myelin sheath - phospholipid-rich material, insulates axon; white appearance
   – formed by oligodendrites (CNS) & Schwann cells (PNS)

 – with unmyelinated axon - AP must travel along entire length of CM
 – with myelinated axon - gaps seen in myelin sheath every 0.1-1.5 mm = **nodes of Ranvier**

saltatory conduction - AP is conducted from 1 node to the next

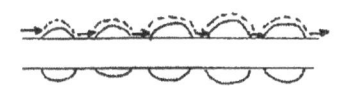

↑ speed of conduction since only CM at nodes undergoes depolarization

– large diameter axons have greater conduction speeds than smaller diameter axons (less resistance)

Type A fibers - large diameter, myelinated, conduction speed = 15-120 m/sec
    e.g., motor neurons; used for rapid response to environment
Type B fibers - medium diameter; lightly-myelinated, conduction speed = 3-15 m/sec
Type C fibers - small diameter, unmyelinated, conduction speed 2m/sec or less
– types B & C found primarily with ANS, where they are used to maintain homeostasis

Organization of nervous tissue -
white matter - bundles of myelinated axons
gray matter - neuron cell bodies + unmyelinated axons
nerve tracts - myelinated axons (white matter) of CNS
nerves - myelinated axons of PNS

within spinal cord - central area is gray matter & outer region is white matter
within brain - gray matter exists in cortex & nuclei and is localized primarily outside of
 deeper white matter

nuclei - cell bodies of CNS   **ganglia** - cell bodies of PNS

within PNS, each axon is surrounded by **endoneurium** - delicate CT wrapping
        perineurium - surrounds groups of axons to form **nerve fascicles**
        epineurium - bind fascicles together to form a nerve

Membrane Potentials

at rest,   +++++++++   due to 1) permeability characteristics of CM
                                    2) presence of - charged proteins within cell
                                    3) Na^+/K^+ ATPase pump

         ─────────
         +++++++++

Na^+ Cl^-
   (K^+)   - large anions too big to move through CM
   (Proteins$^-$) - Cl$^-$ repelled by large anions & attracted to Na$^+$ ∴ stays outside
                   - K$^+$ ~ 100x more permeable than Na$^+$, but K$^+$ attracted to A$^-$ &
                     repelled by Na$^+$ outside
                   - Na$^+$/K$^+$ imbalance maintained & restored by pump
                     (3 Na$^+$ **out**/ 2 K$^+$ **in**)

resting membrane potential (RMP) - due to the above described charge distributions
- at rest, there's a potential difference between inside/outside of CM
- the membrane is **polarized**, RMP = - 85 meV (Table 11.2)

if, [K$^+$] is increased in the ECF, the RMP is less negative → results in hypopolarization
   (the nerve is closer to firing) or depolarization

if, [K$^+$] is decreased in the ECF, the RMP is more negative → results in hyperpolarization
   (the RMP moves further away from depolarization/firing)

if, Na$^+$ membrane permeability increased, Na$^+$ would diffuse into cells → RMP would
   become less negative (i.e., more positive) & depolarization could occur

Na^+/K^+ movement through CM:
1- ion channels allow diffusion through CM (Na^+ moves in / K^+ moves out)
2- Na^+/K^+ ATPase pump moves Na^+ out / K^+ into cell

ion channels - protein pores in CM specific for single ion movement
- permeability of channel affected by electrical charge, 3D structure, size and *gating proteins* (open/close channels)

- diameter of hydrated Na^+ > hydrated K^+
- ∴ K^+ channels are smaller than Na^+ channels; K^+ channels wont accept Na^+

- **voltage-sensitive channels** have gating proteins that open/close in response to changes in electrical charge across CM

- **ligand-gated channels** - chemicals must bind to receptors on the gating protein to open/close channel

In general, Na^+ channels open when membrane depolarizes - open for a short time only

K^+ channels also open with depolarization **but** open more slowly (2nd to open)

w/ depolarization, Na^+ channels open first causing **influx of Na^+** and further depolarization
K^+ channels open later causing **efflux of K^+**, causing CM repolarization

- some Na^+ channels are sensitive to $[Ca^{2+}]$
Na^+ gating protein close channel with ↑ Ca^{2+} in ECF
open channel with ↓ Ca^{2+} in ECF
at normal $[Ca^{2+}]$ in ECF, only a few Na^+ channels are open at any one time

Electrically excitable cells

local potentials - a depolarization event; result of a stimulus applied @ 1 point on cell
- these are examples of **graded potentials** because
magnitude ~ signal strength (Table 11.3)

summation - adding together of local potentials to result in larger depolarization

propagation - spread of local potential over CM
- decreases in magnitude as it spreads over CM away from pt of origin
- can't transfer information over long distances

ACTION POTENTIAL - when local potentials cause depolarization to a level termed the **threshold potential**
- occurs according to an **"all-or-none"** principle (Table 11.4)

AP has a **depolarization phase** - membrane potential moves away from the RMP
- MP becomes more +
- Na^+ channels open during depolarization

repolarization phase - MP returns to the RMP
- MP becomes more -
- Na^+ channels close, K^+ channels open

afterpotential - slight hyperpolarization that occurs with repolarization
- Na^+/K^+ pump reestablishes normal Na^+/K^+ concentrations after AP

AP takes ~ 1-2 msec to complete

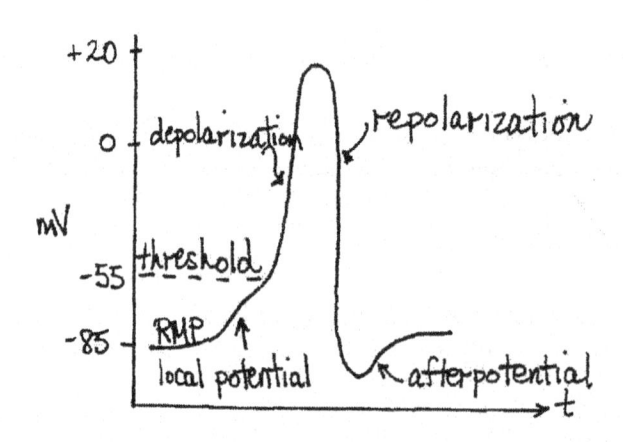

[Diagram: plot of mV vs t showing depolarization rising from RMP (-85) through threshold (-55) to +20, then repolarization back down to afterpotential; local potential arrow at RMP]

refractory period - decreased sensitivity to further stimulation once AP has occurred
- depolarization/repolarization cycle must be completed before 2nd AP generated

absolute refractory period - *complete* insensitivity, if 2nd stimulus occurs between depolarization and beginning of repolarization, no further APs will be generated no matter how strong the 2nd stimulus strength

relative refractory period - after the absolute refractory period has passed, a stronger-than-threshold stimulus can initiate a 2nd AP

propagation of AP - AP spreads to adjacent regions along the CM
- depolarization occurs upstream; repolarization occurs behind travelling AP

AP frequency - strength of stimuli affects the **frequency** of APs
- Subthreshold stimulus → local potential
- Threshold stimulus → local potentials undergo summation and result in a single AP
- Submaximal stimulus – greater than threshold but less than maximal stimulus → results in ↑ AP frequency
- Maximal stimulus (or supramaximal stimulus) → results in AP frequency reaching its maximal rate

AP propagation -
- When threshold is reached, voltage-gated Na^+ channels open in a "domino effect" along the axon, resulting in a wave of depolarization that spreads downstream → propagation of AP
- In myelinated axons, APs "jump" from node to node → saltatory conduction
- In unmyelinated axons, entire length of axolemma must undergo depolarization
- APs travel faster down myelinated, large-diameter axons

Synapse - space over/through which AP travels between cells

   electrical synapse - gap junctions; cells are close enough together to allow direct "jumping" of AP from cell 1 → cell 2
   chemical synapse - **neurotransmitter** released by cell 1 and diffuses across space (synapse), binds to cell 2 receptors & causes depolarization

– stimulus strength determines frequency of APs
– stimulus duration determines how long APs are produced

accommodation - adjustment/return to the RMP even though stimuli continue

Nerve-Nerve Synapse

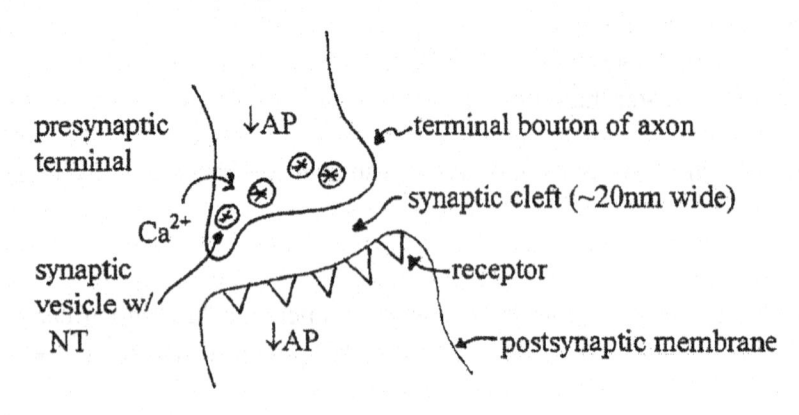

presynaptic terminal — ↓AP — terminal bouton of axon — synaptic cleft (~20nm wide) — Ca²⁺ — synaptic vesicle w/ NT — ↓AP — receptor — postsynaptic membrane

in response to AP along presynaptic axon, Ca^{2+} channels open, causing vesicles to move to cleft and dump NT

— with **Ach** as NT, Achase destroys excess Ach in junction & choline is transported back into presynaptic terminal to reform Ach

— with **norepinephrine** (NE or norepi) as NT, NE is actively transported back into presynaptic terminal where it is inactivated by **monoamine oxidase** (MAO)

— NT also diffuse away from synapse and into ECF
  e.g., NE taken to liver/kidney where catechol-o-methyltransferase inactivates it

— amphetamines ↑ release of NE, block NE reuptake, inhibit MAO → increased stimulation within CNS, increased alertness

— receptors have a high specificity for their NT
— more than 1 type of receptor can exist for a NT
— e.g., NE has 2 receptors ∴ its effect can be either stimulatory or inhibitory depending upon which type of receptor it binds to

neurotransmitters/neuromodulators -
  — some neurons secrete more than 1 NT

neuromodulators - pre/postsynaptically influence the likelihood of an AP being transferred from presynaptic → postsynaptic side of synapse

  e.g., **glutamate** - excitatory NT; with a stroke - ↑ glu release/binding & nitric oxide production resulting in nerve damage
    ? possible to reduce stroke damage by ↓ NO production

EPSPs (excitatory post-synaptic potentials) - when NT & their receptors cause hypopolarization/depolarization

  excitatory neurons - release NT producing EPSPs, 1° due to ↑ Na^+ permeability
  – local anesthetics (novocain) ↓ Na^+ permeability & block EPSPs

IPSPs (inhibitory post-synaptic potentials) - when NT & receptor binding cause hyperpolarization

  inhibitory neurons - release NT producing IPSPs, result from ↑ permeability to Cl^- (moves into cell, makes inside of cell more negative) or
  K^+ (moves out of cell, makes outside more positive)
  – both result in CM moved further away from threshold

axo-axonic synapses - axon of 1 neuron synapses with presynaptic terminal (axon) of another neuron
  – neuromodulators' release at this point influences the [NT] released from the 2nd axons' terminus

  presynaptic inhibition - ↓ NT release; e.g., **endorphins/enkephalins** - by ↓ NT release, they lead to ↓ awareness of pain stimuli

  presynaptic facilitation - ↑ NT release; e.g., **glu & NO** - glu release stimulates NO production & more glu release

summation - combining of local potentials
  – if summation causes local potential > threshold level → AP produced

  spatial summation - 2 APs arrive @ same time @ 2 different presynaptic terminals that synapse with same postsynaptic neuron

  temporal summation - 2 APs arrive close together (in time) at the same presynaptic terminal (1st AP → local potential & before 1st AP repolarizes fully, 2nd AP arrives & summates with remainder of 1st AP to give a depolarization event)

  convergent pathways - many neurons converge synapse with a smaller number of neurons, e.g., motor neurons of spinal cord

  divergent pathways - small number of presynaptic neurons synapse with a larger number of postsynaptic neurons

oscillating circuits - arranged in a circular fashion; prolongs response due to **afterdischarge** (similar to + feedback)
– continues to discharge until fatigued/inhibited

Chapter 12 - The Spinal Cord and Spinal Nerves

Spinal Cord - link between brain & PNS
- runs from foramen magnum → L2 region
- contains enlargements in cervical & lumbosacral regions
- conus medullaris, cauda equina & filum terminale
- CSF fills central canal & subarachnoid space
- protected by 3 meningeal layers: dura mater, arachnoid layer & pia mater

white matter - consists of nerve tracts = **fasciculi**
- myelinated axons in fasciculi carry same type of information
  ascending - sensory ; **descending** - motor
- **funiculi** - columns of white matter

gray matter - consists of nerve cell bodies & dendrites
  dorsal horn - axons of sensory neurons synapse here
  ventral horn - contains cell bodies of motor neurons
  lateral horn - autonomic function

ventral root - carries motor/efferent information away from cord
dorsal root - carries sensory/afferent information; contains **ganglion**

Reflexes - reflex arcs require: 1- sensory receptor
2- afferent/sensory neuron
3- association neuron
4- efferent/motor neuron
5- effector (muscle/gland)

autonomic response - without conscious thought, homeostatic
- remove body from painful stimuli

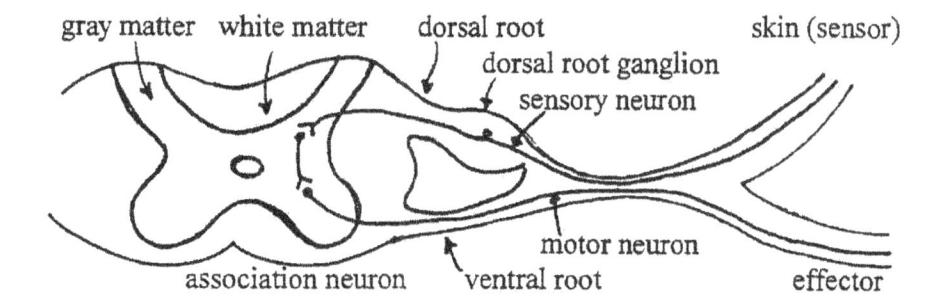

spinal reflexes - **stretch reflex** - simplest, involves afferent & efferent neurons
(no association neuron)
e.g., patellar reflex - muscle spindle = receptor; muscle = effector
– muscle contracts to relieve stretched condition
– **withdrawl reflex** - involves association neurons
eg., pain receptor in skin → sensory neuron → association neuron → motor neuron → effector → contract muscle - withdraw from stimulus
– **crossed extensor reflex** - retains balance during withdrawl reflex

Spinal Nerves - 31 pairs associated with spinal cord
– rootlets combine to form a **ventral** (anterior or efferent) and a **dorsal** (posterior or afferent) root
– spinal nerve = ventral root + dorsal root
– spinal nerve distribution: 8 cervical   C1 - C8
                                  12 thoracic   T1 - T12
                                  5 lumbar   L1 - L5
                                  5 sacral   S1 - S5
                                  1 coccygeal   Co
– spinal nerves have specific cutaneous distributions = **dermatomes**
– spinal nerves branch to form **rami**
  dorsal ramus - supplies muscles and skin of the back
  ventral ramus - form intercostal nerves in thoracic region
                 – the rest of the ventral series join together to form **plexuses**
  sympathetic ramus - supply autonomic nerves

cervical plexus - C1-C4 - neck & head muscle innervation, e.g., phrenic n. → diaphragm

brachial plexus - C5 - T1 - supply upper limb, shoulder, pectoral muscles
                – median n. - damage as it goes through wrist → carpal tunnel syndrome
                – ulnar n. - damaged with blow to elbow
                – axillary n., radial n., musculocutaneous n.

lumbar plexus - L1 - L4 - obturator n. - supplies median thigh
                – femoral n. - supplies anterior thigh

sacral plexus - L4 - S4 - tibial n. - supplies posterior thigh and leg
                – common peroneal (fibular) n. - supplies leg and foot
            – ischiadic n. = sciatic n. = tibial n. + common peroneal n.

coccygeal plexus - S4, S5, Co 1 - supplies pelvic floor and skin over coccyx

Chapter 13 - The Brain and Cranial Nerves

CNS = brain + spinal cord

brain - provides interpretation & integration of sensation, consciousness and cognition

nucleus = nerve cell bodies within CNS, "gray matter"
nerve tracts = bundle of parallel axons within CNS, "white matter"
decussation - crossing over of nerve tracts

Major regions of the brain:

cerebrum - seat of consciousness, perception, thought, control of motor activity
- contains **basal nuclei (ganglia)** - control muscle activity/posture
- includes **limbic system** - autonomic responses, emotion/mood

diencephalon - includes **thalamus** (sensory relay center), **subthalamus** (nerve tracts/ nuclei), **epithalamus** (nuclei for olfaction & **pineal gland**) and **hypothalamus** (maintains homeostasis, endocrine system control)

brainstem - **medulla oblongata** (contains vital centers, brain/spinal cord relay center), **pons varoli** (relay between cerebrum/cerebellum), **midbrain** (visual & auditory reflex centers)

cerebellum - controls muscular coordination/tone

reticular formation - mixing of gray & white matter, occurs in brainstem
- important in sleep/wake cycle

reticular activating system (RAS) - receives visual, auditory & olfactory stimuli to maintain a state of wakefulness

Brainstem - connects brain with spinal cord

    medulla - contains **vital centers**: cardiac center (HR), respiratory center, vasomotor center (BP), swallowing, vomiting, coughing & sneezing center

- **pyramids** - enlargement on the anterior surface of medulla, contains descending (motor) nerve tracts for skeletal muscle control
- **decussation of the pyramids** - nerve tracts experience cross-over, i.e., left side of brain controls motor activity on right side of body & vice versa
- **olives** - 2 oval structures, nuclei involved in balance, coordination, modulation of sound impulses from inner ear
- CN IX, X, XI & XII exit brain at this level

pons - relay center between cerebrum & cerebellum
– CN V, VI, VII, VIII & IX nuclei found here
– contains sleep and respiratory centers

midbrain - tectum = roof; contains 4 nuclei = **corpora quadrigemina**
– **inferior colliculi** - hearing reflex pathway
– **superior colliculi** - visual reflexes - receives input from eye, inferior colliculi, skin & cerebrum; sends fibers to CN III, IV & VI (eyeball movers) & spinal cord (region that innervates neck muscles) → turn head/eye towards stimulus
– **tegmentum** - ascending tracts with paired red nuclei (unconscious regulation/ coordination of motor activities)
– **cerebral peduncles** - descending tracts, major CNS motor pathway
– **substantia nigra** - nuclei contain melanin (∴ dark) - interconnected with basal ganglia

Diencephalon -

thalamus - sensory relay center, contains 2 lobes connected by **intermediate mass**
– **3rd ventricle** - surrounds intermediate mass & separates lobes of thalamus
– most sensory information received here prior to relay to cerebrum
  mediate geniculate nucleus - auditory impulses
  lateral geniculate nucleus - visual impulses
  ventral anterior/lateral nuclei → basal nuclei & motor cortex
  anterior/medial nuclei → limbic system & prefrontal cortex
  ventral posterior nucleus - other sensory information
  lateral dorsal nucleus → cerebral cortex, regulates mood
  lateral posterior nucleus → other thalamic nuclei, sensory integration

hypothalamus - associated with maintenance of homeostasis
– connected via **infundibulum** to **pituitary gland** ∴ exerts control over the endocrine system's "master gland"
– regulates many autonomic functions - HR, BP, GI tract movement, etc.
– controls muscles involved in swallowing and shivering responses
– T regulation - acts as body's thermostat
– regulation of food & water intake - location of hunger/satiety centers
– regulation of sleep-wake cycle
– regulates emotions, site of "mood-altering" drug binding
– contains **mamillary bodies** - provide emotional response to odors
– receives input from visceral organs, taste receptors, limbic system, cutaneous areas associated with sexual arousal & prefrontal cortex (mood regulation)

subthalamus - nuclei associated with basal ganglia system - control motor function

epithalamus - **habenular nuclei** - emotion & smell (visceral response to odors)
  pineal gland - ? function, important in onset of puberty
   – may contain Ca/Mg salts = "brain sand" - used as a landmark for x-ray imaging
   – produces **melatonin** - ↑ at night → sleepy

Cerebellum - controls balance, gross & fine motor control
  – promotes smooth, coordinated movements
  – consists of 2 lateral hemispheres with ridges (folia) & arbor vitae structure
  – **flocculonodular lobe** - balance & muscle tone
  – **vermis** - gross motor control & muscle tone
  – **lateral hemispheres** - fine motor control - smooth/flowing movement
  – field sobriety tests - assess cerebellar function

Cerebrum - left & right hemispheres divided by a longitudinal fissure
  – **gyrus** - folds of tissue, increase surface area
  – **sulcus** - grooves between gyri

frontal lobe - voluntary motor function, motivation, mood, aggression

  central sulcus
   precentral gyrus - primary motor area
   postcentral gyrus - general sensory area

parietal lobe - reception/evaluation of sensory input (except olfaction)

occipital lobe - contains visual centers

temporal lobe - receives olfactory & auditory input; plays a role in memory
   – "psychic cortex" - abstract thought & judgement

insula - deep within lateral fissure, may integrate other areas

cortex - gray matter on outer surface of cerebrum
nuclei - gray matter deep within cerebral hemispheres
cerebral medulla - white matter found between cortex and nuclei of cerebrum
   – consists of 3 types of myelinated nerve tracts:
     association fibers - connect areas within same hemisphere
     commissure fibers - connect R/L hemispheres
       – corpus callosum is the largest of these
     projection fibers - connect cerebrum with other brain regions

meninges - 3 layers of protective CT
    bone → epidural space (cord only) → dura mater → subdural space →
    arachnoid layer → subarachnoid space (with CSF) → pia mater → CNS

CSF (cerebrospinal fluid) - protects & cushions brain/cord
- produced from blood in choroid plexus of each ventricle
- circulates thru ventricles (brain) & central canal (cord)
- CSF leaves subarachnoid space through arachnoid granulations & returns to blood
- hydrocephalus - CSF accumulation/blocked flow

ventricles - cavities within brain lined by ependymal cells
    lateral ventricles - within R/L hemispheres, separated by septum pellucidum
        – CSF passes thru interventricular foramen (Monro) → 3rd ventricle
    3rd ventricle - associated with diencephalon
    cerebral aqueduct - narrow passageway between 3rd & 4th ventricles
    4th ventricle - located between cerebellum & brainstem

3 brain protectors: CSF, dura mater & cranium

Cranial nerves - 12 pair; perform 4 general functions:
1. sensory - special senses (vision, smell); general senses (touch, pain)
2. somatic motor - control of skeletal muscle through motor neurons
3. proprioception - awareness of position
4. parasympathetic - involves regulation of smooth/cardiac muscle & glands

I.   **Olfactory** - sensory - olfaction (smell)

II.   **Optic** - sensory - vision   optic nerve → optic chiasma → optic tract

III.   **Oculomotor** - motor - innervates 4 extrinsic eye muscles (skeletal) & eyelid
        parasympathetic - supplies iris (Δ pupil diameter) and ciliary muscles
        (lens accomodation) of eye - both are smooth muscles

IV.   **Trochlear** - motor - controls extrinsic eye muscle

V.   **Trigeminal** - largest of cranial nerves, contains 3 branches:
    opthalmic (sensory), **maxillary** (sensory), **mandibular** (both)

    sensory - transmits sensations from scalp, nose, upper eyelid (opthalmic);
        palate, jaw, upper jaw (maxillary); lower jaw (mandibular)
    motor - innervates muscles of mastication; palate, throat & middle ear muscles

VI.   **Abducens** - innervates lateral rectus eye muscle

VII. **Facial** - picks up sensory information from the anterior 2/3 of tongue & palate
 – sends motor impulses to facial muscles
 – carries parasympathetic supply to salivary & lacrimal glands

VIII. **Vestibulocochlear (Auditory)** - carries sensory information on hearing and equilibrium

IX. **Glossopharyngeal** - receives taste sensation from posterior 1/3 of tongue
 – supplies throat muscles
 – carries parasympathetic fibers to parotid salivary gland

X. **Vagus** - carries sensory, motor & parasympathetic fibers to/from thoracic & abdominal viscera

XI. **Spinal Accessory** - supplies neck muscles (e.g., trapezius) and pharynx

XII. **Hypoglossal** - supplies tongue and throat muscles

CN III, IV, V, VI, VII, IX, X, XI, and XII also have a proprioception function (sensory)
CN I, II, and VIII are the only purely sensory cranial nerves

Chapter 14 - Integration of Nervous System Function

Sensation and Sensory System

general - widely distributed, no complex sense organ
        e.g., P, T, pain, vibration, proprioception, touch

special - localized w/dedicated organ
        e.g., olfaction, gustation, vision, audition, equilibrium

sensation - perception = conscious awareness

    receptor types:
- **mechano** – e.g., hair cell bending - hearing, equilibrium
- **chemo** - dissolved chemcials; taste & smell
- **photo** - light, retina
- **thermo** - cold, warm; T extremes $\Rightarrow$ pain
- **noci** - pain, free-nerve endings; somatic vs. visceral

    positional types:
- **externo** - cutaneous
- **viscero** - visceral
- **proprio** - joint position

Afferent Nerve Endings:
Merkel's discs - w/in stratum basale; light touch/P
Hair follicle receptors - hair bending
Pacinian corpuscles - dermis/hypodermis; deep P/vibration
Meissner's corpuscles - w/in dermal papillae; 2-pt. discrimination
                  $\uparrow$ # tongue, lips, fingertips; non-hairy surfaces
Ruffini's end organs - w/in dermis, P
Golgi tendon organ - w/in joints, respond to $\uparrow$ tendon tension
Muscle spindles - respond to $\uparrow$ muscle stretch; contraction minimizes damage

Sensory and Motor Nerve Tracts

Spinal pathways - involve primary, secondary & tertiary neurons
    ascending - sensory   **spinothalamic** - lateral - carries pain/T information
                                            anterior - carries touch, P, itch/tickle
                        spinocerebellar - proprioception function

descending - motor   **pyramidal** - muscle tone, skilled movement (conscious)
                                    corticospinal - controls hand movements
                                    corticobulbar - controls face/head movements

                                  75% of motor neurons crossover @ medullary
                                  pyramids, others crossover within cord

                        extrapyramidal - controls unconscious movements
                              rubrospinal - movement coordination
                              vestibulospinal - posture, balance
                              reticulospinal - posture adjustment during movement
                              tectospinal - head/neck movement associated with
                                    visual stimulus

Proprioception - provides information about precise position, rate of movement of
                body parts & range of motion of joints
           – receptors located around joints & muscles

Pain - **somatic** - superficial
     referred - visceral pain "referred" to superficial locations
     phantom - seen with amputation

Sensory and Motor Areas of the Cerebral Cortex

postcentral gyrus - primary somesthetic/general sensory area
- receives pain, T, P information from the thalamus
- **topographical organization** - locates where information from a particular region of the body is processed along the gyrus = **sensory homunculus**
- size of a sensory region reflects the number of sensory receptors in this area; face/hands = ↑ # ; trunk = ↓ #
- **projection** - brain receives sensation but then refers (projects) a sensation to the superficial site at which receptor picked up stimulus

- other primary sensory areas include:
  taste (gustation) area
  olfactory cortex - inferior surface of frontal lobe; conscious/unconscious
     responses to odors are initiated here
  auditory cortex - temporal lobe
  visual cortex - occipital lobe; color/shape/movement processed separately

- association area - adjacent to sensory areas, involved in recognition
  eg., visual association - compares present visual information with past visual
     information; receives input from other brain centers (frontal lobe adds
     emotional value to what is being seen)

precentral gyrus - primary motor cortex, controls voluntary motor activity
- **topographical organization** - largest areas of motor cortex represent muscle groups that have the most motor units present
- **motor homunculus** - areas with small motor units (fine control) are larger on the homunculus - mouth & hands

premotor area - organizes/stages motor activities; sends information to motor cortex
- **apraxia** - hesitancy in performing complex/skilled motor activities due to problems with premotor area

prefrontal area - an association area; largest in humans
– involved with motivation, provides foresight to plan movements
– important in regulation of emotional behavior/mood
– **prefrontal lobotomy** - removes aggressive behavior/causes personality changes

Speech - most individuals are left-side speech dominant

Wernicke's area - sensory speech, located within parietal lobe
– comprehends & formulates coherent speech
Broca's area - motor speech, located within inferior part of the frontal lobe
– receives information from Wernicke's area
– sends information to premotor & 1° motor areas to initiate muscle movements necessary for speech production

right cerebral hemisphere - controls muscle activity & receives sensory input from the left side of the body
– involved with spatial perception & musical ability, 'artsy'
left cerebral hemisphere - controls muscle activity & receives sensory input from the right side of the body
– 'analytical' abilities, math and speech abilities

Brain waves - **EEG (electroencephalogram)** - measures electrical activity of brain

alpha (α) - awake & quiet state, recorded with eyes closed
beta (β) - present with increased mental activity
theta (θ) - seen in children & adults with frustration, ↓ with socialization
delta (δ) - pattern seen during deep sleep state

Memory - 3 types: **sensory** - very-short term, less than 1 sec
— operative during sensory input evaluation
short-term - few sec → few min; usually limited to units of 7
— only so much can be stored - old information wiped out by new input
long-term - may involve physical changes to neurons
— long-term potentiation

— create a **memory trace (memory engram)**
— **declarative memory** - retention of facts, localized in hippocampus & amygdaloid nucleus (adds emotional overtones)
— **procedural/reflexive memory** - development of skills
— information stored in cerebellum & premotor cortex
— includes conditioned reflexes

Basal nuclei (ganglia) - nuclei located in inferior cerebrum, diencephalon & midbrain
— play a role in organizing/planning & coordinating movement/posture
— have inhibitory effect on muscle activity
— decrease muscle tone, eliminate unwanted muscular activity

include: **subthalamic nucleus** - within diencephalon
substantia nigra - within midbrain
corpus striatum - deep within cerebrum
— **caudate nucleus**
— **lentiform nucleus** - putamen & globus pallidus
dyskinesia - disorders associated with the basal nuclei
e.g., cerebral palsy, St. Vitus dance, Parkinson's, nystagmus

Limbic system - controls visceral function through ANS
— includes parts of cerebral cortex (paleocortex) & diencephalon
eg, cingulate gyrus, hippocampus, basal nuclei, thalamus, hypothalamus, mamillary bodies, olfactory bulbs/cortex, fornix

Chapter 15 - The Special Senses

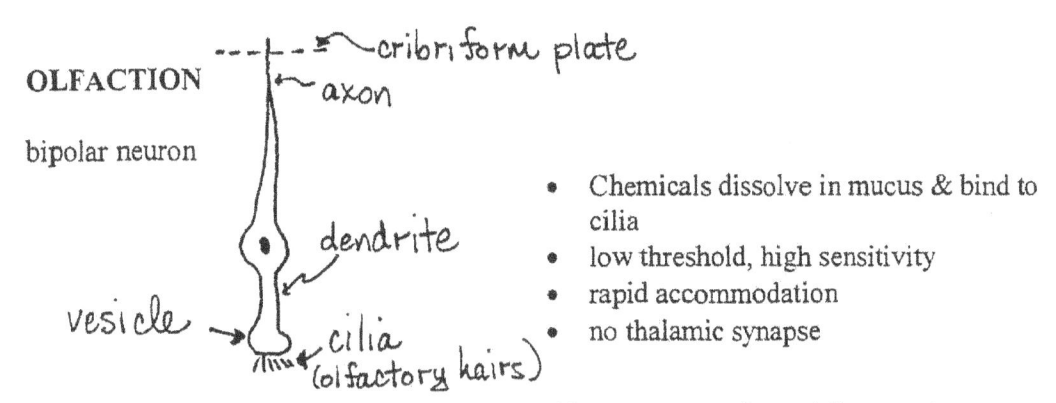

OLFACTION

bipolar neuron

- Chemicals dissolve in mucus & bind to cilia
- low threshold, high sensitivity
- rapid accommodation
- no thalamic synapse

CN I → mitral cell → association neuron → olfactory cortex (lateral fissure of cerebrum)

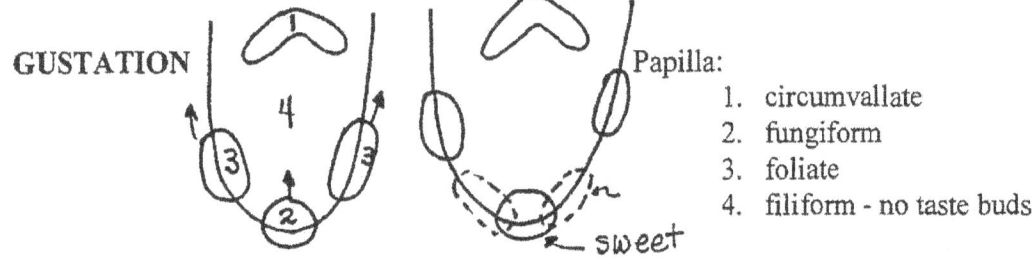

GUSTATION

Papilla:
1. circumvallate
2. fungiform
3. foliate
4. filiform - no taste buds

taste buds = support + gustatory cells (hairs or microvilli)
chemicals bind to receptor on gustatory hair → depolarization
CN VII, IX, X → medulla → thalamus → gustatory cortex (postcentral gyrus)
80% of taste is actually due to olfaction
basic types of taste – sour, salty, bitter, sweet, umami (savory)

VISION

eye:
1. **Fibrous tunic**: **sclera** - protective, extrinsic muscles attach
   cornea - transparent, refraction, avascular

2. **Vascular tunic**: **choroid layer** - melanin (absorbs stray light)
   intrinsic eye muscles (smooth) = { **ciliary body** - lens accomodation
   iris - regulates pupil size
      circular - constriction (↑ light, close), parasym
      radial - dilation (↓ light, far), sympathetic

3. **Retina: pigmented**
   nervous: - photoreceptors (rods & cones)
               - bipolar neurons
               - ganglionic neurons

Pathway of AP to brain:
rod/cones → bipolar neurons → ganglionic neurons → CN II → optic chiasma → optic tract → thalamus (lateral geniculate nucleus) + superior colliculi (midbrain) → visual cortex (occipital lobe)

optic disk - blindspot - lacks photoreceptors, exit of CN II

anterior compartment of eye: anterior chamber,   cornea → iris
                                      posterior chamber,   iris → lens

aqueous humor - ~ intraocular P, provides nutrients, refraction
                    - continually made/ drained - canal of Schlemm
                    - blocked drainage, ↑ P ⇒ glaucoma

posterior compartment of eye: behind lens

vitreous humor - ~ intraocular P, eye shape, refraction
                 - presses retina against back wall of eye

lens: biconvex   - anchored by suspensory ligaments
                             - attached to ciliary body
                             - bends, magnifies, inverts image

lens accomodation: "flat" - for distance vision (ciliary muscles relax)
                          "plump" - for close/near vision (ciliary muscle contract)

refraction - bending of light rays passing through different medium

        cornea → aqueous humor → lens → vitreous humor

focal point - point at which light rays converge & cross
    - inverted image focused on retina at some point beyond focal point

- emmetropia - true/normal
- myopia - "near"
- hyperopia - "far"
- presbyopia - "old"

reflection - light rays bounce off a nontransparent surface
        - reduced by pigmented surfaces of eye (melanin)

near point of vision - closest an object can approach & stay in sharp focus
-any closer → blurring; 2-3" (child); 4-6" (young adult); 20" (~ 45 yr.); 60" (~ 80 yr.)

convergence - as object approaches, eyes rotate medially
(use medial rectus muscle ⇒ "lazy eye")

visual field - everything that can be seen with only 1 eye open is in the visual field of that eye (lateral/medial component)

binocular vision - overlap of R/L visual fields ⇒ <u>depth perception</u>

RODS - responsible for vision in low-light (night vision), high sensitivity
- black & white vision (no color discrimination)
- concentrated @ periphery of retina (peripheral vision)
- contain rhodopsin as visual pigment

rhodopsin $\xrightarrow{hv}$ retinal + opsin ["bleaching of rhodopsin"]
　　　　　　　pigment (vit A)　protein
　　　　　⇒ <u>hyperpolarization</u> of rod CM

↑ light, rhodopsin bleached; eyes less sensitive to light
↓ light, more rhodopsin made; retina now more light sensitive (↑ pigment)
- provides vision @ low light intensities

CONES - color vision; sharpness/visual acuity
- ↑ # cones @ fovea centralis (within macula lutea)
- pigment = iodopsin; requires ↑er light intensities to activate
- 3 types of cones: blue, green, red

AUDITION = hearing

external ear - auricle → external auditory canal → tympanic membrane (eardrum)

middle ear - malleus → incus → stapes → oval window (on vestibule)
- this area normally filled w/air

auditory ossicles - transmit/amplify vibrations

auditory (Eustachian) tube - equalizes P on both sides of eardrum
- opens to pharynx

inner ear - **cochlea** - hearing
vestibular apparatus - **vestibule** = utricle + saccule (static equil.)
- **semicircular canals** (dynamic equilibrium)

cochlea –

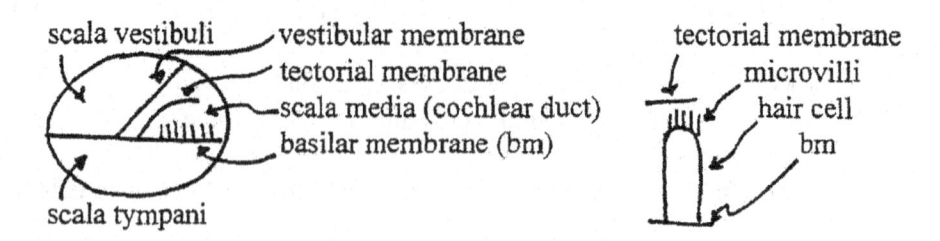

scala vestibuli — vestibular membrane
— tectorial membrane
— scala media (cochlear duct)
— basilar membrane (bm)
scala tympani

tectorial membrane
microvilli
hair cell
bm

organ of Corti = hair cells + tectorial membrane
(spiral organ)
cochlea is fluid-filled - scala media - contains endolymph
  scala tympani/vestibuli - contains perilymph

Pathway of AP to brain:
vibration of fluid within inner ear → bend hair cells → depolarization (AP) → CN VIII → medulla → midbrain (inferior colliculi) → thalamus (medial geniculate nucleus) → auditory cortex (temporal lobe)

Table 15.2 – steps involved in hearing

    pitch - frequency/repetition rate of vibrations
- ↑ frequency - proximal basilar membrane vibrates
- ↓ frequency - distal basilar membrane vibrates
- humans hear 20 - 20,000 Hz (physiological range)
    1,000 - 4,000 Hz (optimal range)
    loudness - amplitude/height of waves
- 0+ decibels; sounds over 125 dB are painful

EQUILIBRIUM

    static balance - associated with static labyrinth
- evaluates position of head relative to gravity
- processes information on linear movement
- macula with hair cells; $CaCO_3$ (otoliths) sit atop hair cells
- otolith movement deflects cilia → depolarization

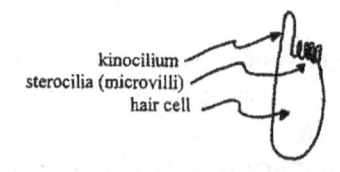

kinocilium
sterocilia (microvilli)
hair cell

dynamic/kinetic balance - associated with semicircular canals
- ampulla contains crista (with hair cells)
- above crista is a gelatinous cap (cupula)
- cupula is displaced by endolymph fluid movement within semicircular canals; hair cells bend → depolarization
- fluid movement is in direction opposite that of physical movement

anosmia - lack of olfaction
methylmercaptan - added to natural gas
emmetropia - image in focus on retina; 20/20
myopia - "near"; image focused in front of retina, corrected with concave lens
hyperopia - "far"; image focused behind retina; corrected with convex lens
presbyopia - with age; ↓ lens accommodation due to ↓ lens flexibility
bifocals - contain both distance & close-up corrections

glaucoma - ↑ P in anterior compartment of eye
astigmatism - due to scratch/aberration of cornea and/or lens
cataract - lens becomes "cloudy"
color-blindness - genetic lack of cones, x-linked, 1^0 affects males
night blindness - rod dysfunction
macular degeneration - loss of sharp, central visual field
neonatal gonorrheal opthalmia - $AgNO_3$, antibiotics

conduction deafness - blocked sound transmission
sensorineural deafness - nerve pathway damage
tinnitus - ringing in ears
otitis media - middle ear infection
motion sickness - over-stimulation of semicircular canals
space sickness - under stimulation of semicircular canals

Chapter 16 - The Autonomic Nervous System

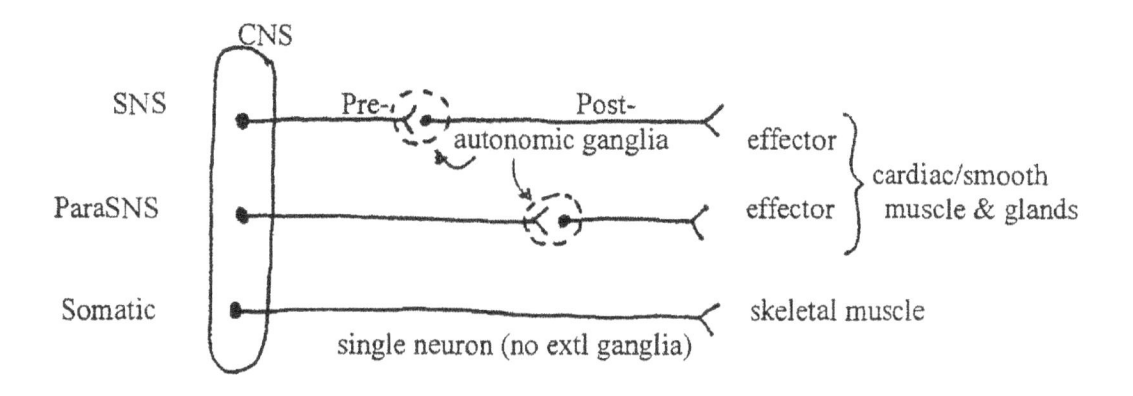

```
              CNS
           ┌──────┐
   SNS     │  ●───┼── Pre─ ·◌·  Post─ ───┤ effector ┐
           │      │    autonomic ganglia              │  cardiac/smooth
   ParaSNS │  ●───┼─────────────── ◌ ─────┤ effector ┘  muscle & glands
           │      │
   Somatic │  ●───┼────────────────────────┤ skeletal muscle
           └──────┘  single neuron (no extl ganglia)
```

ANS - possesses 2 neurons from CNS → effector

Sympathetic division - short preganglionic/long postganglionic neurons
- sympathetic ganglia close to spinal cord ("chain ganglia")

Parasympathetic division - long preganglionic/short postganglionic neurons
- "terminal ganglia" near effector

SYMPATHETIC - "fight-or-flight"; prepares body for stress
- adrenergic system, NT = norepinephrine
- thoracolumbar division
- effects are diffuse and generalized
- axons exit sympathetic ganglia via spinal nerves, sympathetic nerves, splanchnic nerves or adrenal gland

PARASYMPATHETIC - "feeding and breathing", vegetative activities
- restores homeostasis
- craniosacral division
- vagus nerve (CN X) contains 75% of all paraSNS neurons
- effects are specific & localized
- cholinergic system, NT = acetylcholine
- **SLUDD**: salivation, lacrimation, urination, digestion, defecation

most organs receive dual innervation (both SNS & paraSNS) with possible outcomes = **antagonistic, cooperative** or **complementary**

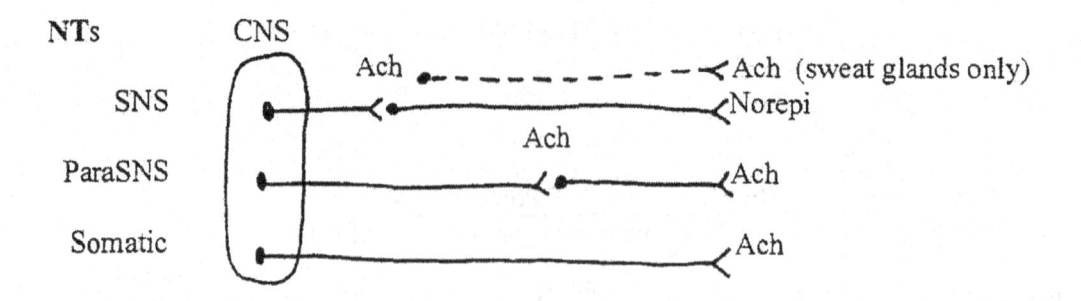

```
NTs        CNS
                    Ach •----------< Ach (sweat glands only)
    SNS    •----<•-----------------< Norepi
                         Ach
    ParaSNS •----------<•----------< Ach

    Somatic •-------------------------< Ach
```

Ach - used by all preganglionic fibers; all postganglionic paraSNS & all somatic fibers
Norepi - used by all postganglionic SNS fibers (exception - sweat glands use Ach)

Receptors - binding of NT to receptor can cause +/- effects

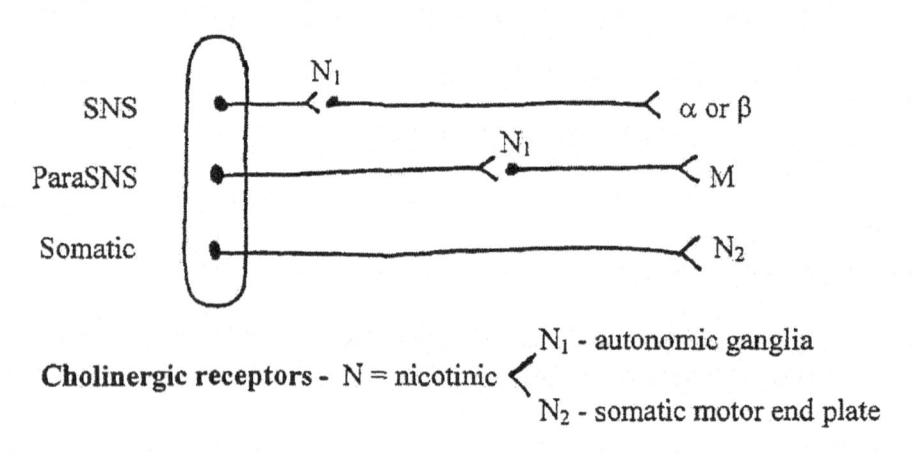

```
              N₁
    SNS    •--<•----------------< α or β
                      N₁
    ParaSNS •--------<•---------< M

    Somatic •--------------------< N₂
```

Cholinergic receptors - N = nicotinic
- N_1 - autonomic ganglia
- N_2 - somatic motor end plate

M = muscarinic - at paraSNS effector cells

Adrenergic receptors - α_1 & α_2 ; β_1 & β_2 ; both α/β can be +/- in effect

Ach binding to N is always stimulatory; Ach binding to M can be either +/-
Epi (adrenal medulla) binds to same α,β receptors as norepi

Drug therapies –
Agonist for receptor = affinity + efficacy (action) ≡ stimulatory agent
Antagonist for receptor = affinity without efficacy ≡ blocking agent

   Cholinergic antagonist = antimuscarinic/nicotinics = parasympatholytics

   Cholinergic agonists (mimic Ach) = parasympathomimetics

   Adrenergic agonists (mimic norepi) = sympathomimetics

   Adrenergic antagonists = sympatholytics, eg. β-blockers

Chapter 17 - Functional Organization of the Endocrine System

- Nervous and endocrine systems are the control systems of the body.
- Help maintain homeostasis; under negative feedback control.

Hormones - "chemical messengers", produced by ductless (endocrine) glands
- travel through the bloodstream to target cells
- influence all aspects of cell metabolism/function

	Nervous system	Endocrine system
Control chemicals		
Cells affected		
Time course		
Duration		
Distance traveled		
Secretory cells		
Regulation		

Types of intracellular messengers: NT; hormones; neurohormones; neuromodulators; paracrines; autocrines; pheromones

Chemical nature of hormones: proteins -
   aa derivatives (tyr) -
   steroids -
   FA derivatives -

Mechanisms affecting hormone secretion: [metabolite] -
   nervous system -
   another hormone (tropic) -

Hormone transport - free or bound to a plasma protein

endocrine gland →hormone → bloodstream (w/ or w/o plasma protein) →
pass out of vessel (thru pore or diffuse thru wall) → interstitial spaces →
target cell (binds to CM or passes thru CM)

Half-life of hormone prolonged by binding to plasma protein or addition of a CHO group
   half-life = length of time it takes for elimination of half a dose of a substance from circulation

Removal of hormones via: excretion -
   metabolism -

   conjugation -

   active transport into cells -

Hormone receptors: 1. membrane-bound receptors (protein hormones) -
                   2. intracellular receptors (steroid hormones) -

1. Membrane-bound receptors - hormone (= 1st messenger) binds to CM receptor → production of another chemical within cell (= 2nd messenger); rapid acting

G protein - located on inner surface of CM; contains α, β, γ subunits
                    1- alter CM permeability
   α-GTP and: 2- alter phosphorylase activity
                    3- ↑/↓ [2nd messenger]

Cascade effect - few 2nd messenger molecules → large effect (amplification)

Example 1: G protein opens CM Ca^{+2} channels; Ca^{+2} acts as 2nd messenger
   Epinephrine binds to β-adrenergic receptors of heart → ↑ Ca^{+2} permeability →
   ↑ $[Ca^{+2}]_{intracellular}$ → ↑ force/rate of cardiac contractions

Example 2: G protein ↑ [cAMP, cGMP]; cAMP/cGMP act as 2nd messengers
   Glucagon binds to liver CM → α-GTP produced → ↑ adenylyl cyclase activity →
   cAMP → ↑ protein kinase activity → activates enzymes of glycogenolysis

Example 3:
   Epinephrine binds to smooth muscle cells → α-GTP produced →
   DAG (diacylglycerol) and IP$_3$ (inositol triphosphate) produced →
   release of Ca^{+2} from ER (IP$_3$) and ↑ PG synthesis and phosphokinase activity (DAG)
   → ↑ smooth muscle contraction

2. Intracellular receptors:
- Steroid hormones pass thru CM to bind to cytoplasmic or nuclear receptors; slow acting
- Time required to "turn on genes" (results in mRNA and protein synthesis)
- Latent period = time it takes hormone to bind to receptor, then cause response

Chapter 18 - The Endocrine Glands

Regulatory functions of the endocrine system:

1- metabolism and tissue maturation
2- ion regulation
3- water balance
4- immune system regulation
5- heart rate and blood pressure regulation
6- control of blood glucose and other nutrients
7- control of reproductive functions
8- uterine contractions and milk release

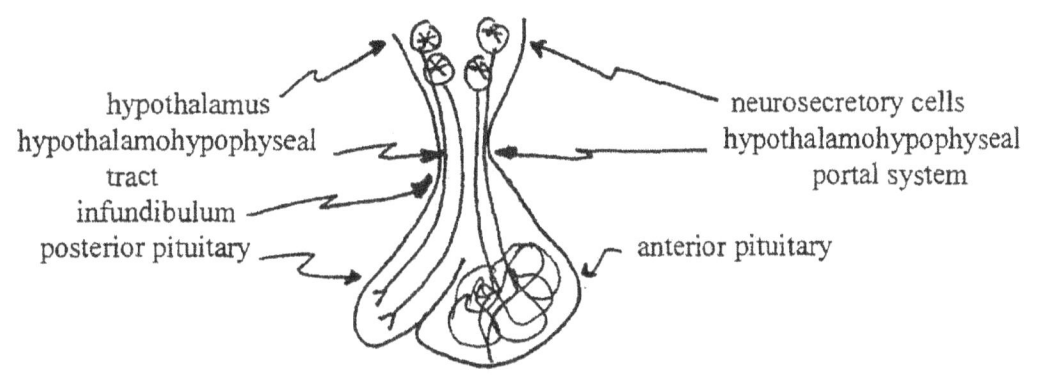

Labels: hypothalamus, hypothalamohypophyseal tract, infundibulum, posterior pituitary, neurosecretory cells, hypothalamohypophyseal portal system, anterior pituitary

hypothalamus - bridge between NS → endocrine system
- important in the maintenance of homeostasis
- produces ADH & oxytocin
- produces various RH/IH neurohormones that regulate the activity of the anterior pituitary

posterior pituitary (neurohypophysis) - secretes/releases ADH & oxytocin, which are manufactured in hypothalamus

anterior pituitary (adenohypophysis) - "master gland", produces hormones that regulate various body functions & other endocrine glands

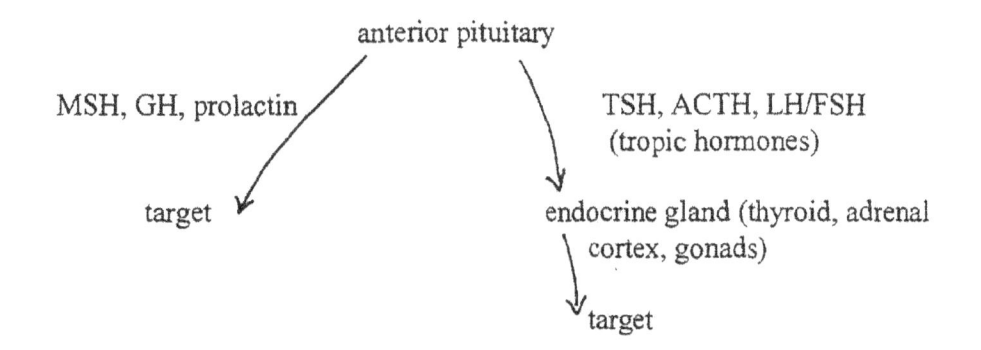

anterior pituitary

MSH, GH, prolactin → target

TSH, ACTH, LH/FSH (tropic hormones) → endocrine gland (thyroid, adrenal cortex, gonads) → target

oxytocin - from paraventricular nuclei of hypothalamus
hypothalamus → posterior pituitary → uterine smooth muscle
→ mammary glands, "let-down"
- under positive feedback control

ADH (vasopressin, antidiuretic hormone) - from supraoptic nuclei of hypothalamus
hypothalamus → posterior pituitary → kidney tubules → ↑ H_2O reabsorption
↓
produced in response to ↑ Osm, ↓ BP   ↓ urine vol.
↓ ADH = diabetes insipidus → ↑ urine volume   ↑ blood vol.
           ↓ blood volume, ↓ BP   ↑ BP
↓ ADH production/action with EtOH, caffeine   ↓ Osm

HYPOTHALAMUS	ANTERIOR PITUITARY	TARGET
GHRH, GHIH	GH	all cells → anabolism
PRH, PIH	Prolactin	mammary glands → milk
GnRH	LH, FSH	gonads → sex hormones, etc.
TRH	TSH	thyroid → T_3/T_4
CRH	ACTH	adrenal cortex → cortisol
	MSH	skin → produces melanin
	lipotropins	adipose → ↑ fat breakdown
	β - endorphins	act as natl. opiates, ↑ w/stress, exercise

Growth Hormone (GH) = somatotropin
- 1^0 anabolic hormone, stimulates growth in most tissues
GHRH → GH → ↑ aa uptake into cells → ↑ protein synthesis → tissue growth
→ ↑ fat breakdown (FA used as fuel source for anabolism)
→ spares glucose usage → ↑ glycogen synthesis (liver)

↓ [GH] child → dwarfism (short stature, normal body proportions,
(hyposecretion)   normal mental development)

↑ [GH] child → gigantism ; ↑ [GH] adult → acromegaly

pygmy - [GH] normal, genetically lack ability of liver to produce <u>somatomedins</u>
(synergist required for GH action).

↑ [aa]
↓ [blood glucose] } ↑ **GH**   GH peaks @ night during sleep
stress, exercise

THYROID GLAND

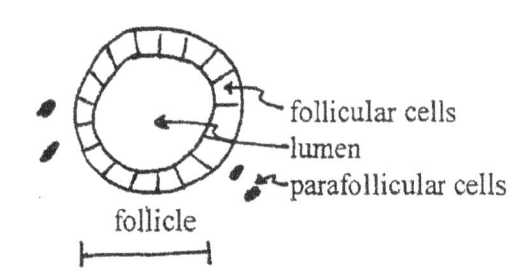

parafollicular cells → **calcitonin** → bone → stabilizes osteoclasts   ↓ blood
↑ osteoblast activity Ca^{2+}

follicular cells - TRH → TSH → thyroid follicle → **T_3/ T_4**
      T_3 = triodothyronine; T_4 = thyroxine
   - I^- pumped into follicle
   - thyroglobulin (contains tyr) synthesized within follicle cells
   - I^- → I; I added to tyr of thyroglobulin
   - thyroglobulin with I-tyr moved to lumen
   - T_3/T_4 formed/stored in lumen (2 week supply, bound to thyroglobulin)
   - thyroglobulin passes back into follicle cells
   - T_3/T_4 released into interstitial spaces
   - T_3/T_4 travel in blood bound to TBG (thyroxine-binding globulin)
   - 40% T_4 converted to T_3, T_3 more potent/ 1° hormone @ target cells

T_3 passes thru CM → binds to intracellular receptors → ↑ protein synthesis
        → binds to mitochondria → ↑ ATP, heat production

1° effect of T_3/T_4 → ↑ metabolic rate; synergist for GH

hyposecretion: nutritional - lack of I, causes goiter
               cretinism - in children
               myxedema - in adults

hypersecretion: goiter (hypertrophy)
                Graves's disease

PARATHYROID GLANDS → PTH (parathyroid hormone)

↓ Ca^{2+} → ↑ **PTH** → bone → ↑ osteoclast activity → bone resorption → ↑ Ca^{2+} / ⓟ
                     kidney → ↑Ca^{2+} reabsorption / ↑ ⓟ excretion
                     intestine → ↑ Ca^{2+} / ⓟ absorption in intestine (vitamin D)

overall result of PTH → ↑ blood Ca^{2+} / ↓ blood
PTH & calcitonin are antagonists of one another

↑ PTH or ↓ calcitonin → hypercalcemia
↓ PTH or ↑ calcitonin → hypocalcemia

ADRENAL GLANDS   <   medulla → epinephrine/norepinephrine

                         cortex →   zona glomerulosa → aldosterone
                                    zona fasciculata → glucocorticoids/cortisol
                                    zona reticularis → androgens

adrenal medulla   - component of ANS, innervated by sympathetic nerves
                             - produces 80% epinephrine / 20% norepinephrine
                             - promotes "fight or flight" reactions
      epinephrine → ↑ [blood glucose], ↑ glycogenolysis (liver), ↑ fat breakdown
                     → ↑ heart rate / force of contraction
                     → dilates blood vessels to skeletal & cardiac muscle
                     → constricts blood vessels to skin, kidney, GI tract
    ↑ excitement, injury, stress, exercise, ↓ [blood sugar] ⇒ ↑ epinephrine

adrenal cortex - steroid hormones

     aldosterone → kidney tubules , → ↑ Na^+ reabsorption, ↑ H_2O reabsorption
                                   → ↑ blood volume / ↑ BP
                                   → ↑ K^+ / H^+ excretion → alkalosis, ↑ pH / ↓ [H^+]

              ↓ BP, ↑ K^+ / ↑ H^+ ⇒ ↑ aldosterone prod.

     cortisol - 1° glucocorticoid
              - ↑ fat breakdown, ↓ glucose/aa uptake in skeletal muscle
              - ↑ protein breakdown
              - ↑ gluconeogenesis (aa → sugar); ↑ blood glucose
              - ↑ intensity of inflammatory response (↓ # WBC, ↓ histamine release)

↑ stress, hypoglycemia → ↑ CRH → ↑ ACTH → ↑ cortisol
Addison's disease   - ↓ aldosterone, cortisol
Cushing's syndrome - ↑ aldosterone, cortisol, androgens → "moon face",
                                       fat redistribution, ↑ [blood glucose]
Adrenal diabetes   - hypersecretion of cortisol

PANCREAS
- **exocrine - acinar cells produce digestive enzymes**
- **endocrine - islets of Langerhans secrete hormones**
  - α (alpha) - produce glucagon } antagonists
  - β (beta) - produce insulin
  - δ (delta) - produce somatostatin

insulin - antagonist of glucagon
    - ↑ [blood sugar] → insulin → ↓ [blood sugar]
    - with insulin, ↑ glucose, aa uptake into cells
    - 1° targets = liver, skeletal muscle, fat
    - overall, an anabolic hormone: glucose → glycogen; aa → protein

insulin shock - with ↑↑ insulin, ↓↓ glucose, CNS malfunctions
                 - with ↓↓ insulin, ↑↑ glucose, (but, glucose doesn't enter cells!)

glucagon - antagonist of insulin
    - ↓ [blood sugar] → glucagon → ↑ [blood sugar]
    - 1° target = liver; causes glycogenolysis, gluconeogenesis

secretion control -
+ with hypoglycemia - ↓ insulin/ ↑ glucagon → ↑ blood sugar to normal
+ with parasympathetic stimulation of GI tract → ↑ insulin/ ↓ glucagon
+ with sympathetic stimulation of GI tract → ↓ insulin/ ↑ glucagon
+ GI hormones (secretin, CCK, gastrin) → ↑ insulin
+ somatostatins → ↓ insulin & glucagon

- immediately after meal → ↑ glucose → ↑ insulin, ↓ epi, glucagon, GH, & cortisol
                        → also ↑ parasympathetic stimulation

- hours after meal → ↓ glucose → ↓ insulin, ↑ epi, glucagon, GH & cortisol

diabetes mellitus - Type I = juvenile, "early onset", insulin-dependent, little/no insulin production, ? autoimmune

                    Type II = "adult onset", noninsulin dependent,
                          inability of insulin-receptors to respond, linked to obesity, genetic-predisposition

gestational diabetes - temporary condition during pregnancy, ↓ sensitivity to insulin

pineal gland: produces melatonin ⇒ ↓ GnRH → ↓ reproduction fcns.

thymus gland: produces thymosin → regulates development of immune system

Chapter 19 - The Blood

Functions: transportation
protection
maintenance of homeostasis - T, pH, hemostasis

Blood - CT with a liquid matrix = **plasma**
4-5 L (females) → 5-6 L (males)   ~ 8% body weight

whole blood = formed elements - erythrocytes (RBCs)
– leukocytes (WBCs)
– thrombocytes (platelets)
– plasma - a colloidal suspension = whole blood - formed elements

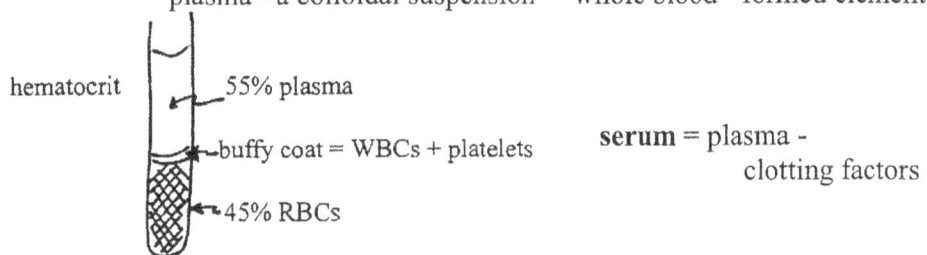

hematocrit — 55% plasma
buffy coat = WBCs + platelets
45% RBCs

serum = plasma - clotting factors

Plasma - 91% water, 9% suspended materials, "straw" color

plasma proteins - albumin - stays within vessel, draws water to it
– helps maintain osmotic concentration of blood
globulins - used as transport molecules (α, β) and antibodies (γ)
fibrinogens - used in blood clotting

hematopoeisis - formation of blood cells from hematocytoblasts in bone marrow
erythropoeisis - RBC production

RBCs - ~ 5×10^6/cc
– primary function is O_2/CO_2 transport
– secondary function is production of HCO_3^- (bicarbonate ion)/regulates blood pH

within RBC, $CO_2 + H_2O$ <u>carbonic anhydrase</u> $H_2CO_3 \rightleftharpoons H^+ + HCO_3^-$

RBC shape = **biconcave disc**; flexible, **anucleate** in circulation
contains **hemoglobin, Hb** - O_2 transport protein
[Hb] = 12 - 18 g/100ml blood

Hb consists of 〈 **globin** - 4 protein chains; 2 α, 2 β

heme - Fe-porphoryn pigment, 1 per chain

Fe binds O$_2$ ∴ Hb when fully saturated carries 4 O$_2$

HbO (oxyhemoglobin) - carries O$_2$ (bright red)
HbCO$_2$ (carbaminohemoglobin) - carries CO$_2$
HbCO (carboxyhemoglobin) - carries CO; interferes with O$_2$ transport
 — 5-15% total Hb (higher in smokers)
HHb (reduced hemoglobin) - carries H$^+$; low pH form
deoxyhemoglobin - without O$_2$ (brick red/blue)
fetal Hb - greater affinity for O$_2$; must accomplish gas exchange across placenta

erythropoeisis - takes ~ 4 days
 < adults - red marrow - flat bones, femur, humerus
   fetus - all long bones + liver & spleen

hematocytoblast → proerythroblast → intermediate (polychromatic) erythroblast
                                                                            ↓
       erythrocyte ←—margination— reticulocyte ← late erythroblast

erythropoeisis requires: Fe for Hb synthesis; Fe absorption added by vitamin C
 — aa for globin synthesis
 — folic acid, vitamin B$_{12}$ for DNA synthesis
   (prerequisite for protein synthesis)

erythropoeitin - hormone from kidney, released when kidney blood [O$_2$] is low
testosterone - stimulates erythropoeitin production (∴ males have more RBCs)

RBC life span (~100-120 days) - short because without nucleus, cytoplasmic machinery
 can't replace proteins

hemolysis - RBC destruction; worn-out RBCs removed from circulation by macrophages
 of liver (kupffer cells)

 Hb recycling - aa of globin reused to make new proteins
  — Fe (attached to ferritin) brought from liver → bone marrow
  — heme group - converted from biliverdin → bilirubin
   — secreted with bile from liver
  jaundice - accumulation of bilirubin; UV light decomposes bilirubin

Blood Typing - based on Ab/Ag agglutination reaction

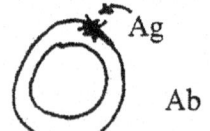

Ag on RBC membrane (A & B types)

Ab circulates in plasma (AbA, AbB)

ABO Blood Groups

Blood Type	Ag (RBC)	Ab (plasma)	donate to*	receive from *
A	AgA	AbB	A, AB	A, O
B	AgB	AbA	B, AB	B, O
AB	AgA, AgB	-	AB	A, B, AB, O
O	-	AbA, AbB	A, B, AB, O	O

AB = universal recipient   O = universal donor
*misleading - always best to give someone their own blood type
  e.g., if you give type O → type AB individual - type O blood contains AbA/AbB that can attack host's RBC Ags

transfusion reaction - results from **mismatched** blood
  – Ab/Ag reaction (agglutination) causes RBCs to **clump** together; hemolysis follows

Rh Blood Group - Antigen D, "Rh factor" found on RBC CM
  – Ab to Rh factor - not found within plasma of Rh⁻ person <u>unless</u> exposed to Rh⁺ blood

Rh⁺	Rh Ag	no Ab
Rh⁻	no Ag	no Ab

exposure to Rh+ blood → develop Ab

erythroblastosis fetalis = hemolytic disease of the newborn
  Rh⁻ mom, Rh⁺ fetus; mom's immune system makes Ab to Rh factor/Ag on baby's RBC
  RhoGam - suppresses mom's immune system, prevents sensitization

Leukocytes - WBCs, 5,000 - 9,000/cc
  – defensive cells; phagocytic; nucleated
  – found outside blood vessels within tissue spaces
  – **diapedesis** - movement of WBCs out of blood vessel
  – WBCs travel ameoboid movement
  – movement directed by **chemotaxis**

1- **granulocytes** - lobed nuclei, granules within cytoplasm

   neutrophils - most common WBC (60-70%)
     – #s increase during bacterial infection/decrease during viral infection
     – 1ˢᵗ WBC at site of an infection
     – motile; phagocytic; granules = lysosomes
     – polymorphonuclear neutrophils (PMNs) - nuclei vary in shape/# lobes

eosinophil - (1-4%); bright red granules in cytoplasm
- reduce inflammation by releasing antihistamines
- seen during allergic reactions & in parasitic infections

basophils - rarest of WBCs (0.5-1%); promote inflammation
- large blue granules release: **heparin** - anticoagulant
  histamine - vasodilator, ↑ inflammation

2- **agranulocytes** - simple nucleus, granules absent

monocytes - (2-6%); largest WBC
- leave circulation to become **macrophages** (very phagocytic cells)
- 2nd WBC at site of infection
- seen in increased numbers during chronic infections

lymphocytes - (20-30%), smallest WBC
  T lymphocytes - participate in cell-mediate immunity
  B lymphocytes - produce antibodies

Platelets - Thrombocytes - 300,000/cc
- represent fragments of **megakaryocte**
- play key role in **hemostasis** (prevent blood loss)
  1- formation of **platelet plug** (seals hole in small vessel)
  2- formation of **clots** (seals off larger tears in vessels)

PGs - promote **platelet aggregation**; aspirin - prevent PG release
aspirin - used in heart attack/stroke prevention **but** can cause hemorrhage at childbirth

Hemostasis - prevention of blood loss

1- **vasoconstriction** - immediate/temporary closure of vessel
  - **vascular spasm** (SNS) & **vasoconstricting chemicals** (released by platelets)

2- **platelet plug** - platelets activated when in contact with exposed CT (collagen fibers)
  - produce ADP, making surfaces "sticky"
  - activates other platelets causing them to attach to plug
  - forms "framework" for clot

3- **coagulation** - clot formation; follows a **cascade** of events

```
        extrinsic pathway                    intrinsic pathway
  (damaged tissue release tissue factor,   (starts with factor XII binding to
         TF, thromboplastin)                 collagen & activated platelets)
                         ↘         ↙
   I.        Prothrombin activator formed (from factor X)
                              ↓
   II.       prothrombin  prothrombinase  thrombin
                              ↓
   III.      fibrinogen (soluble)  thrombin  fibrin (insoluble)
```

- fibrin threads polymerize forming basis of blood clot
- many clotting factors require Ca^{2+} or vitamin K (from bacterial metabolism in colon)
- deficiency of Ca^{2+}/vitamin K can cause bleeding disorders

anticoagulants - prevent blood clotting; antithrombin, heparin, prostacyclin (PG)
clot retraction - squeeze out serum; pulls wound closed
fibrinolysis - clot dissolution after healing; **plasmin** action
thrombus - stationary clot;   **embolism** - dislodged/moving clot

Blood disorders:
   anemia - ↓ #RBCs, ↓ [Hb]

   aplastic anemia - bone marrow disorder

   iron-deficiency anemia - ↓ [Fe]

   pernicious anemia - lack of vitamin B_{12} and/or intrinsic factor
      intrinsic factor (from stomach) is required for vitamin B_{12} absorption

   sickle cell anemia - genetic defect, abnormal HbS
      RBCs collapse/sickle with ↓ O_2 & clog capillaries; fragile RBCs rupture

   hemophilia - lack of clotting factors
      Type A - lack of factor VIII; Type B - lack of factor IX; Type C - lack of factor XI

   leukopenia - ↓ #WBCs; **leukocytosis** - ↑ #WBCs; **leukemia** - ↑↑ #WBCs

   thalessemia - abnormal Hb

   polycythemia - ↑ #RBCs; 2° - due to ↑ altitude, ↓ O_2 conditions (need for ↑ #RBCs)

   septicemia - "blood poisoning" with microorganisms

   infectious mononucleosis - infects lymphocytes

Chapter 20 - The Heart

Heart - a pump, enclosed in a fluid-filled pericardial sac
    fibrous pericardium
    serous pericardium - parietal vs. visceral layers

3 Layers of Heart: Epicardium (visceral pericardium)
                          Myocardium (muscle)
                          Endocardium (inside lining)

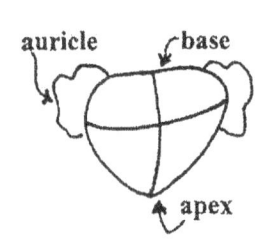

2 **atria** - filling/receiving chambers
2 **ventricles** - pumping chambers

auricles - outer, flaplike extensions of atria
        - ↑ surface area; pectinate muscle

Valves prevent backflow of blood:
    Atrioventricular valves (AV)   **tricuspid** (right)
                                       bicuspid (mitral) (left)

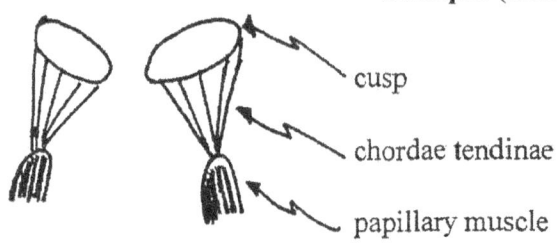

cusp

chordae tendinae

papillary muscle

Semilunar valves (SL):   **pulmonary** (right w/in pulmonary artery)
                              aortic (left - w/in aorta)

interatrial septum - site of fossa ovalis; fetal foramen ovale
                           – R/L atrial shunt

R atrium	L atrium
R ventricle	L ventricle

interventricular septum

R /L coronary arteries - supply oxygenated blood to heart
    R - posterior interventricular, R marginal
    L - anterior interventricular, circumflex, L marginal

small cardiac vein → great cardiac vein → coronary sinus → R atrium

Arteries - carry blood **away** from heart
Veins - carry blood **towards** heart

    Angina pectoris - referred pain; due to ↓ O_2 (ischemia) to heart muscle

– blocked coronary a. can lead to **myocardial infarction** (MI = heart attack); due to prolonged lack of O_2

    angioplasty - balloon compresses plaque against vessel wall
    coronary bypass - reroute blood around occluded vessel
    enzyme treatment - dissolves clots; minimizes MI damage
        e.g., streptokinase, urokinase, plaminogen activator

FLOW OF BLOOD THROUGH THE HEART

```
                Tricuspid      Pulmonary SL      blood picks up O₂                        Bicuspid        Aortic SL
                 valve            valve          drops off CO₂                             valve           valve
                   ⇓                ⇓                  ⇓                                     ⇓               ⇓
1. IVC
2. SVC ──→ Rt. Atrium ──→ Rt. Ventricle ──→ Pulmonary A. ──→ LUNGS ──→ Pulmonary V. ──→ Lft. Atrium ──→ Lft. Ventricle ──→ Aorta
                                                            4.                          5.                                    │
                                                                                                                              │
                                                                                                                              ↓
                                                                                                                              6.
         3. Coronary                                                                                    Lft. & Rt.
            sinus                                                                                       coronary
              ↑                                      unloads O₂                                         arteries
              │                                      picks up CO₂                                           │
              │                                          ⇑                                                  │
              └──────────────────────────────────── CAPILLARIES ←────────── To Body ─────────────────────────┘

          DEOXYGENATED BLOOD                          GAS EXCHANGE                    OXYGENATED BLOOD
```

Veins - blood **toward** the heart
Arteries - blood **away** from the heart

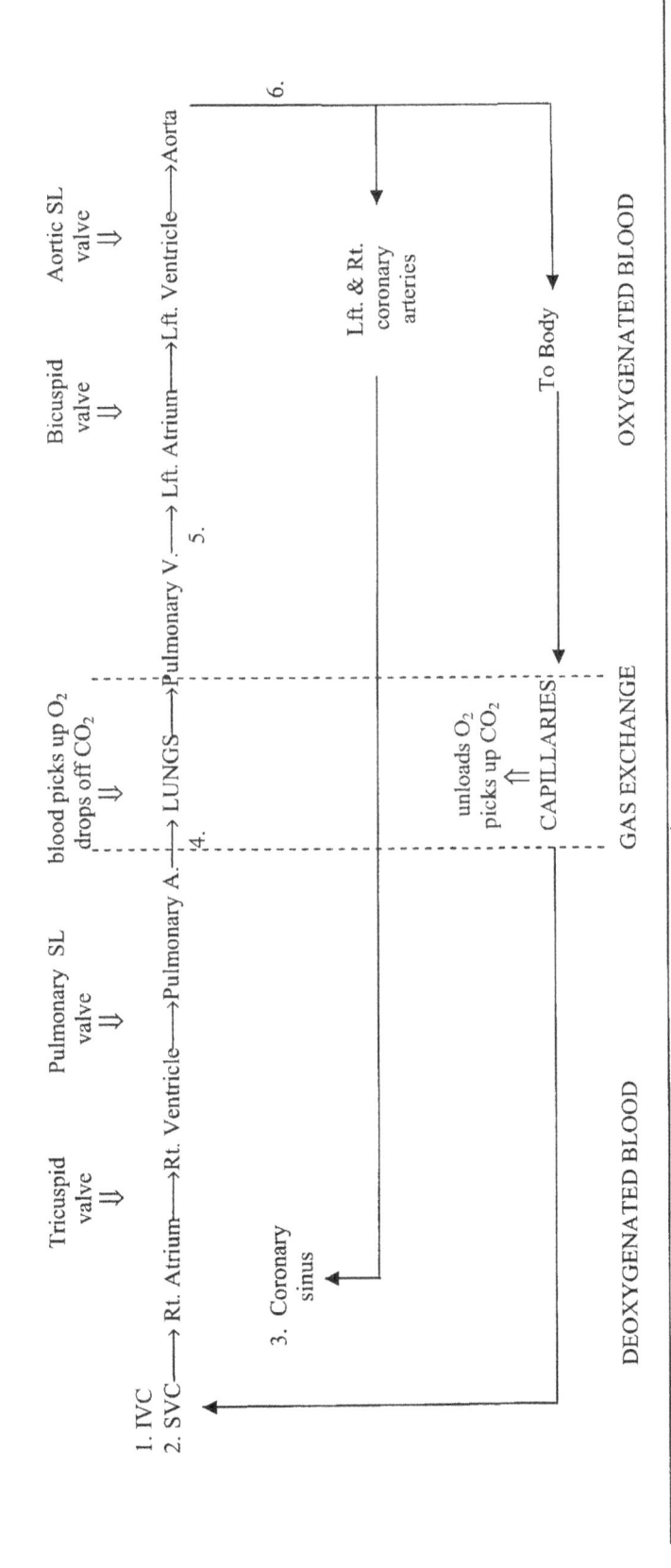

1, 2, 3 - carry deoxygenated blood back from system circulation
4 - deoxygenated blood → lungs
5 - oxygenated blood → heart
6 - carries O₂ - rich blood to systemic circulation

Cardiac muscle cells - "excitable", produce AP
- no motor nerve supply; contract w/o direct innervation
  <u>intercalated disc</u> - gap jcn.; allow rapid communication between cardiac muscle cells; helps synchronize beating

electrical conduction system - produces **autorhythmicity**

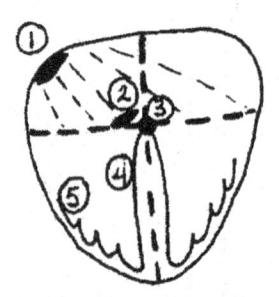

1. **sinoatrial node (SA)**: "pacemaker"
   - establishes cardiac rhythm (beats/min)
   - initiates AP that spreads through atria
   - "fires" ~ every 0.8 sec. → 70 beats/min

2. **atrioventricular node (AV)**: "delay switch"
   - holds up AP for ~ 0.1 sec.
   - ensures atria complete contraction before AP spreads to ventricles

3. **(AV) atrioventricular bundle (Bundle of His)**: at top of interventricular septum

4. **right/left bundle branches**: run down either side of interventricular septum

5. **Purkinje fibers**: spread AP to ventricular muscle

   *** ventricular contraction begins at the apex and rises towards the base ***

<u>Ectopic foci/ pacemaker</u> - with heart block, some other conduction component will take over the role of the SA node

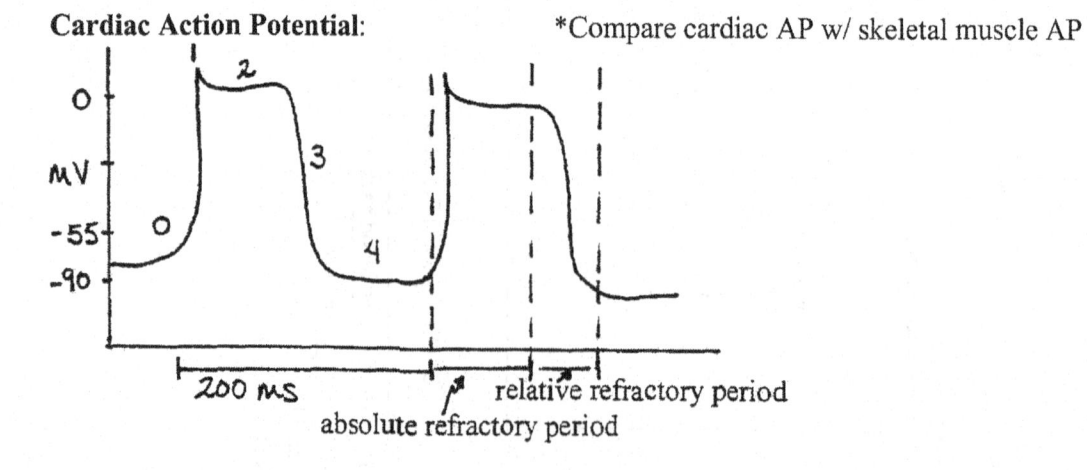

Cardiac Action Potential: *Compare cardiac AP w/ skeletal muscle AP

200 ms
absolute refractory period
relative refractory period

Phase ∅ - region of rapid depolarization (due to opening of fast Na^+ channels)
- Na^+ enters cells → depolarization

Phase 1 - partial repolarization (due to closing of Na^+/ opening of K^+ channels)

Phase 2 - plateau stage (due to opening of slow Ca^{2+} channels/ closing of K^+ channels)
- responsible for prolonged refractory period

Phase 3 - repolarization; closing of Ca^{2+}/ opening of K^+ channels

Phase 4 - RMP reestablished

--- long refractory period ensures relaxation is complete before another AP is initiated ---
** prevents tetany in heart muscle **

tetrodotoxin - blocks Na^+ fast channels
Mn^{2+}/verapamil - blocks Ca^{2+} slow channels; reduces autorhythmicity, ↓ HR
epi/norepi - ↑ HR by influencing Ca^{2+} channels
- reduces refractory period so heart depolarizes again

Electrocardiogram - ECG - electrical events; summation of APs in heart muscle

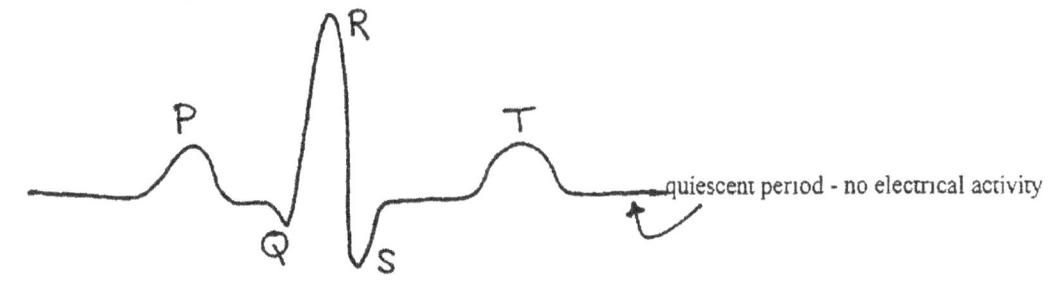

time between 2 P waves = **cardiac cycle** (~0.8 sec.)

P wave - SA node fires giving atrial depolarization (atrial systole then follows)
QRS complex - atrial repolarization (hidden)
- ventricular depolarization (ventricular systole follows)

T wave - ventricular repolarization

region between P/QRS represents firing of AV bundle, bundle branches & Purkinje fibers

ECG abnormalities are important in clinical diagnostics

cardiac arrhythmias -
    tachycardia - ↑ HR, > 100 beats/min
        - caused by ↑ T, excessive sympathetic stimulation
    bradycardia - ↓ HR, < 60 beats/min
        - caused by ↑ vagal stimulation, athletic training, carotid sinus syndrome
    atrial flutter - 300 P waves/ min; synchronous activity, ectopic APs w/in atria
    fibrillation - asynchronous contractions, ↓ pumping efficiency
    defibrillation - electrical shock stops asynchronous activity;
        - opportunity for SA node to reestablish normal rhythm
    SA node block - no P wave; AV node can serve as ectopic pacemaker, ↓ HR
    AV node block - interrupts signal delivery to ventricles
        - ventricles must develop ectopic foci; $1^{st} \rightarrow 2^{nd} \rightarrow$ complete
    PVCs - prolonged QRS complex, inverted T wave; ectopic foci w/in ventricles
    PACs - prolonged P wave; due to ↓ sleep, ↑ coffee/smoking
    Pericarditis - inflammation of serous pericardium
    cardiac tamponade - ↑ fluid w/in pericardial cavity
        - fluid compresses heart; can't relax fully, ↓ filling

Cardiac Cycle - repetitive pumping process that occurs every ~ 0.8 sec. (Table 20.2)

Systole = contraction
Diastole = relaxation

- all 4 chambers can be in diastole 2 same time = quiescent period

- when 1 set of chambers are in systole, the other set of chambers must be in diastole

Ventricular filling - volume w/in ventricle results from: (for left ventricle)
    <u>passive filling</u> - ~ 100 ml, during atrial diastole
    <u>active filling</u> - ~ 20 - 30 ml, during atrial systole
    end diastolic volume = active + passive filling ≅ 120-130 ml

Ventricular contraction - pressure increases as ventricles enter systole
    - when P(ventricle) > P(atrium) → AV valves close

    isometric (isovolumic) contraction - pressure ↑ but volume remains constant
    - both AV & SL valves closed ∴ blood can **not** exit

    ejection - when P(ventricle) > P(aorta) → SL valves open
        - blood is ejected from the ventricle

    end systolic volume = 50-60 ml, blood left in ventricle after systole

Stroke Volume = amount of blood ejected from ventricle with each contraction
= end diastolic volume - end systolic volume = 120 ml - 50 ml = 70 ml/beat

ventricular relaxation - as ventricle enters diastole P(ventricle) decreases

when P(ventricle) < P(aorta) → SL valve closes

isometric (isovolumic) relaxation - both SL & AV valves are closed

when P(ventricle) < P(atrium) → AV opens & ventricular filling begins again

** when P curves cross, valves between chambers/ chambers & arteries open or close **

begin & end each cardiac cycle with AVopen/ SL closed

ventricular filling	AV open/ SL closed
isometric (isovolumic) contraction	AV closed/ SL closed
ejection	AV closed/ SL open
isometric (isovolumic) relaxation	AV closed/ SL closed

* most of cardiac cycle is spent in diastole
* atria spend more time in diastole than ventricles
* ventricles spend more time in systole than atria

Heart Sounds - due to valve closure
  1st sound - "lubb" - AV valves close
      - occurs at beginning of ventricular systole
  2nd sound - "dupp" - SL valves close
      - occurs at beginning of ventricular diastole
  3rd sound - faint sound due to blood flowing through ventricles
      - seen in young, thin individuals; easier to hear if individual is lying down

murmurs - gurgle/swish, abnormal sounds due to leaky valves, backflow of blood
  stenosis - narrowed valve, never fully opens
      - turbulent blood flow; rushing/whining sound

** ventricular systole is approximately the time between the 1st & 2nd heart sounds **

Cardiac Output (CO) = amount of blood pumped out of heart per min
= heart rate x stroke volume   (**HR x SV**)

with HR = 72 beats/ min & SV = 70 ml/ beat:
resting CO = 72 beats/min x 70 ml/beat ≅ 5,040 ml/min or 5 L/min
maximum CO = 120 beats/min x 200 ml/beat ≅ 24 L/min
cardiac reserve = maximum CO - resting CO = 24 L/min - 5 L/min = 19 L/min

Blood Pressure (BP) = CO x PR (peripheral resistance)

PR ≡ is total resistance against which blood must be pumped

ANYTHING THAT ALTERS CO or PR WILL ALTER BP

CO	PR	
changes **HR**	Viscosity of blood	↑ viscosity, ↑ **PR**
changes **SV**	radius of vessel	↑ radius, ↓ **PR**
	length of vessel	↑ length, ↑ **PR**

REGULATION OF THE HEART

intrinsic regulation - results from normal functional characteristics of the heart itself

venous return - amount of blood returned to the right atrium

Starling's Law - more the heart fills (stretches), the greater the strength of contraction
 ↑ preload = ↑ stretching of ventricular walls ⇒ ↑ CO
 ↑ preload causes ↑ SV
 ↑ venous return → ↑ end diastolic vol. ⇒ ↑ SV ⇒ ↑ CO
 ↑ venous return ⇒ stretching of SA node → ↑ rate AP ⇒ ↑ HR

afterload - pressure ventricles must produce to overcome P(aorta)
 - P(aorta) must exceed 170 mmHg before interfering with ventricle function
 (i.e., heart is more sensitive to preload than afterload)

extrinsic regulation - hormonal or neural control mechanisms

1. **neural control** - cardioregulatory center in medulla
 - vagus nerve delivers parasympathetic stimulation to the heart

 ** parasympathetic → Ach → ↓ HR **

 - cardiac nerve delivers sympathetic stimulation to the heart

 ** sympathetic → norepi → ↑ HR & ↑ force of contraction → ↑ SV **

2. **hormonal control** - takes longer to act than neural control

 ** adrenal medulla → epi → ↑ HR & ↑ force of contract → ↑ SV **

 sympathetic NS/adrenal medulla stimulated by stress, excitement, activity

 SNS ⇒ ↑ CO by 50-100%   ParaSNS ⇒ ↓ CO by 10-20%

Bainbridge reflex - works with Starlings law of the heart

↑ stretch receptors → ↑ AP to medulla → ↑ sympathetic stimulation ⇒ ↑ CO, ↑ HR
(right atrium)

Heart & Homeostasis -

1- Effect of BP on the heart -

Baroreceptors - stretch receptors, ↑BP ⇒ ↑ stretch ⇒ ↑ AP produced
– located in internal carotid a & aorta

afferent fibers - send information to cardioregulatory center of medulla
efferent fibers - lead from cardioregulatory center → heart

↑ BP (detected by baroreceptors) → ↓ SNS / ↑ ParaSNS ⇒ ↓ HR & SV ⇒ ↓ BP
↓ BP → ↓ ParaSNS / ↑ SNS ⇒ ↑ HR & SV ⇒ ↑ BP

2- Effect of pH/CO_2 on the heart -

↑CO_2 (↓ pH) (medullary chemoreceptors) → ↑SNS / ↓ ParaSNS ⇒ ↑ HR & ↑ SV
↑ blood delivery to lungs to remove excess CO_2 ∴ pH returns to normal

3- Effect of ions on the heart -

↑ K^+ ⇒ ↓ HR, ↓ SV; with 2x [K^+] → heart block
↑ Ca^{2+} ⇒ ↑ force of contraction & usually ↓ HR
↓ Ca^{2+} ⇒ ↓ force of contraction & ↑ HR

4- Effect of T on the heart -

↓ body temperature ⇒ ↓ HR   ↑ body temperature ⇒ ↑ HR

Chapter 21 - Peripheral Circulation & Regulation

Cardiovascular system = "closed" circulatory system
   Pulmonary circulation = right ventricle → pulmonary a → lungs
                left atrium ← pulmonary v ←

Systemic circulation = left ventricle → aorta → elastic arteries → muscular arteries
                                                                            ↓
               right atrium ← vena cava ← veins ← venules ← capillaries

in <u>systemic</u> circulation (includes coronary circulation) - arteries carry O_2-rich blood,
             veins carry deoxygenated blood
in <u>pulmonary</u> circulation - arteries carry deoxygenated blood, veins carry O_2-rich blood

Capillaries - important as site of gas exchange
    - contain only endothelium (1-cell layer in thickness; simple squamous epithelium)
Types of capillaries - classified according to pore size:
   1. **fenestrated** - contain pores (fenestrae = windows)
                - found in highly permeable tissues, eg. kidney, intestines
   2. **sinusoidal** - very large pore diameter; found in endocrine tissue
   3. **continuous** - without fenestrae → less permeable
                - found in muscle/nervous tissue

       "<u>sinusoids</u>" - large diameter sinusoidal capillaries
               - large molecules & (sometimes) cells pass through
               - found in liver & bone marrow

       "<u>venous sinuses</u>" - similar to sinusoidal capillaries, found in spleen

       <u>thoroughfare channel</u> - blood flows from metarteriole → venule
               - continuous blood flow; no exchange

       <u>true capillaries</u> - opening regulated by **precapillary sphincter**
               - blood flow is slow & intermittent
               - exchange takes place

Materials move out of capillaries & into tissues:
       + through fenestrae
       + through spaces/gaps between endothelial cells
       + through CM

   - lipid soluble materials (O_2 & CO_2) diffuse through CM
   - polar materials must pass through fenestrae &/or gaps between cells

Blood vessel wall - 3 layers (tunics)

1. **tunica intima** (interna) - innermost, endothelium (simple squamous epithelium)
    - lines lumen, in contact with blood
    - some smooth muscle/elastic fibers present
    - only layer found in capillaries

2. **tunica media** - contains smooth muscle & elastic tissue
    - 1°/thickest layer in arteries

3. **tunica adventitia** (externa) - outermost layer, contains fibrous CT
    - thickest layer seen in veins

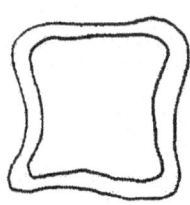

artery
smaller diameter
thicker wall
withstands P

vein
larger diameter
thinner, irregular wall
"blood reservoirs"

arterioles - smooth muscle constriction/dilation → regulates blood delivery & BP

veins - 1-way valves prevent backflow
   varicosity - valves are stretched, veins dilate

arteriovenous anastomoses: blood flows directly from arteries → veins (AV shunt)
   (e.g., sole of foot, palm, fingertips, nailbeds → T regulation)

innervation of blood vessels -
- arterioles receive sympathetic input/no parasympathetic supply
- ↑ sympathetic → constriction
- some sympathetic firing present at all times → "tone"
- ↓ sympathetic → dilation

Arteriosclerosis - loss of arterial elasticity with age
   - "hardening" of arteries, results in ↑ BP
   - thickening of tunica intima, elastic fibers "change"
   - when fat accumulates between elastic/collagen fibers →
      produce lesions that protrude into lumen → ↓ blood flow, ↑ resistance
      ⇒ ↑ work of heart

Atherosclerosis - deposition of atheromatous ("fatty") plaque within arterial wall
   - plaque contains cholesterol → later calcifies

Lymphatic vessels - return system (have valves to prevent backflow)
lymph capillaries → lymphatics → lymph nodes → lymphatics →
→ right lymphatic duct → right subclavian v.
→ thoracic duct → left subclavian v.

Physics of Circulation

Factors that influence circulation - physical characteristics of blood

Viscosity - resistance of a liquid to flow
↑ viscosity ⇒ ↑ pressure required for blood to flow through vessel
↑ viscosity - 1° due ↑ RBCs/↑ hematocrit
↑ viscosity ⇒ ↑ resistance to flow ⇒ ↑ BP

Flow patterns -
1. **Laminar** flow - seen in small-diameter vessels with "streamlined" smooth walls 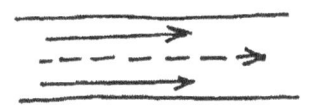
   - fluid at center of vessel experiences ↓ resistance
   - flow is faster within center region of vessel
   - fluid next to vessel wall encounters more resistance ∴ flow is slower at wall

2. **Turbulent** flow - seen in larger vessels, flow disrupted due to constriction/curve; "audible"

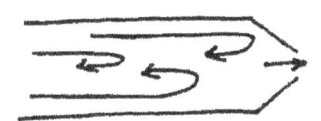

BP - force exerted by blood against vessel wall, measured in mm Hg
- measured by listening for Korotkoff sounds due to turbulent flow
- **BP cuff -** increase P to cut off flow to brachial artery
  Sphygmomanometer = BP cuff attached to a manometer
  - with ↑ P, occluded artery, no blood flow, no sounds
  - with gradual P release, hear tapping sounds associated with turbulent blood flow through partially occluded artery

1st Korotkoff sound = systolic P
disappearance of Korotkoff sounds = diastolic P (return to laminar flow)

Blood always flow from high P → low P
- greater Δ P → greater flow rate
- arteries have highest P ∴ velocity of blood flow greater in arteries

Resistance opposes flow -   ↑ R ⇒ ↓ flow;   ↓ R ⇒ ↑ flow

Vessel diameter influences resistance -
$\uparrow$ diameter, $\downarrow$ R $\Rightarrow$ $\uparrow$ flow; $\downarrow$ diameter, $\uparrow$ R $\Rightarrow$ $\downarrow$ flow
(arteries)   (capillaries - important for gas exchange)

Vessel length influences resistance - $\uparrow$ length $\to$ $\uparrow$ R $\to$ $\downarrow$ flow

Poiseuille's Law - Flow $= \dfrac{P_1 - P_2}{R} = \dfrac{P_1 - P_2}{8vl/r^4} = \dfrac{(P_1 - P_2)r^4}{8vl}$

flow $\approx$ 4th power of vessel radius
eg. $\uparrow$ vessel radius by a factor of 2x, blood flow $\uparrow$ by 16x

with exercise - dilate vessels to skeletal muscles ($\downarrow$ sympathetic outflow)
- leads to $\downarrow$ R $\to$ $\uparrow$ blood flow to active tissue

Law of LaPlace - Force = Diameter X Pressure
- force that stretches vascular wall is $\approx$ D X P
- explains critical closure P

Crictical Closure P - minimum P required to keep vessel open
- when P drops below this point $\to$ vessel collapse $\to$ flow stops $\to$
$\to$ **circulatory shock**

aneurysms - bulges that occur in weakened blood vessel walls
- result from $\uparrow$ force being applied to weakened vessel walls

vascular compliance - $\uparrow$ BP results in $\uparrow$ vessel volume
- easier a vessel wall stretches, the greater its compliance will be

veins - largest compliance, account for 64% of blood volume, act as "blood reservoirs"
- compliance of veins ~ 24x greater than that of arteries

% Blood Volume Distribution
7% - coronary circulation
9% - pulmonary circulation
84% - systemic circulation = (64% in veins, 15% in arteries, 5% in capillaries)

Δ P within systemic circulation:
average P(aorta) = 100 mmHg
medium size artery = 95 mmHg
small artery = 85 mmHg
arterioles = 30 mmHg   } greatest Δ P experienced in transition from
venules = 10 mmHg   artery $\to$ arterioles
vena cava = 0 mmHg

* Greatest ΔR occurs within arterioles; also see greatest ↓ flow rate here

* Constriction/dilation of arterioles (SNS) provides most of BP control

Pulse pressure = systolic P - diastolic P = 120 - 80 = 40 mmHg
                - influenced by SV & vascular compliance

hypertension = high BP; borderline vs. clinical
hypotension = low BP

Capillary Dynamics - nutrient/waste product exchange
      - simple diffusion down concentration gradient
      - nutrients move out of capillary & into interstitial spaces (eg. O_2)
      - waste products move out of interstitial spaces & into capillary (eg. CO_2)

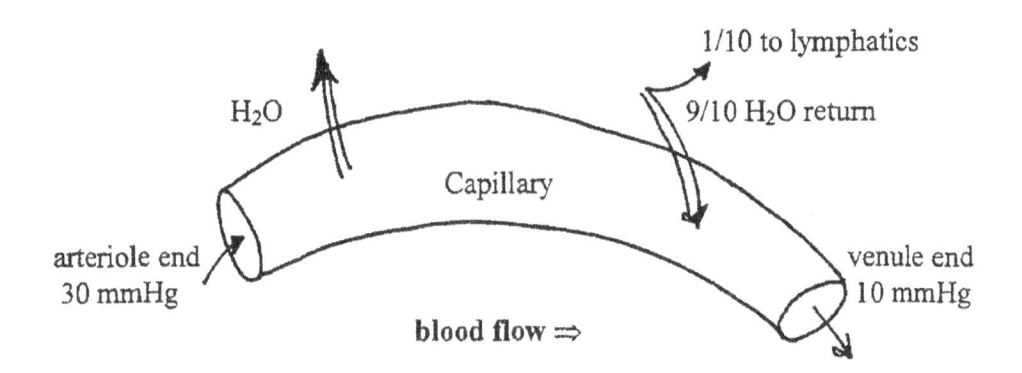

[Diagram of capillary: arteriole end 30 mmHg, venule end 10 mmHg, blood flow →; H_2O leaving at arteriole end; 1/10 to lymphatics, 9/10 H_2O return at venule end]

net movement of H_2O out of capillary at arteriole end due to capillary hydrostatic P (CHP)

net movement of H_2O back into capillary at venule end due to colloidal osmotic P (COP)

    COP - H_2O drawn to large solutes (i.e. proteins) left within capillaries

Edema due to   1. ↓ [plasma proteins], ↓ COP, ↑ fluid accumulation in tissue spaces
                2. ↑ capillary permeability, plasma proteins leak to interstitial spaces
                3. ↑ BP, ↑ CHP, ↑ H_2O outward movement at arteriole end

Gravity - greatly elevates capillary pressures; forces more fluid out of capillaries
Veins -   ↑ venous tone → ↑ venous return → ↑ SV → ↑ CO
    - pooling of blood within veins ↓ VR → ↓ CO

structural/physiological aids for veins - valves, skeletal muscle contraction

Control of Blood Flow to Tissues

1. **Local control of flow** - due to metabolic need
    - low O_2, high CO_2 - signals need for ↑ delivery
    - brain, kidney, liver always supplied fully regardless of [O_2, CO_2]
    - blood flow through capillaries to tissues regulated by local factors:
        - vasodilator substance produced by tissues with ↑ metabolism
        - with ↑ CO_2, ↑ H^+, ↑ lactic acid or ↓ nutrients (O_2, glucose)
            precapillary sphincters dilate ⇒ ↑ flow

2. **Nervous system control** - sympathetic division of ANS
    - vasomotor center within medulla has sympathetic outflow only

    vasomotor tone - SNS outflow leads to continual partial state of contraction of blood vessel smooth muscle

    ↑ SNS ⇒ CONSTRICTION;   ↓ SNS ⇒ DILATION
    with exercise, ↑ SNS @ skin, viscera; ↓ SNS @ skeletal/cardiac muscle

3. **Humoral control** - via hormones & vasoactive chemicals, reinforces NS activity

Mean Arterial Pressure (MAP) = CO x PR = (HR x SV) x PR

↑ HR, ↑ SV, ↑ PR (↑ viscosity, ↑ length, ↓ radius) ⇒ ↑ MAP
↓ HR, ↓ SV, ↓ PR (↓ viscosity, ↓ length, ↑ radius) ⇒ ↓ MAP

Short-Term Regulation of BP -
    fast-acting mechanisms; control BP for periods of seconds to minutes

1. **baroreceptors** - sensitive to stretch, located within carotid a. & aortic arch

    Baroreceptor reflex stimulated:

    with ↑ BP → ↑ stretch → ↑ stimulation of vasomotor/cardioregulatory center
        ↑ parasympathetic → (vagus n. → heart) → ↓ HR → ↓ CO ⇒ ↓ BP
        ↓ sympathetic ⟶ dilate arterioles (↑ r) → ↓ PR ⇒ ↓ BP

    Baroreceptor reflex inhibited:

    with ↓ BP → ↑ sympathetic → ↑ HR & ↑ SV → ↑ CO ⇒ ↑ BP
    (↓ stretch)   → ↓ r (vasoconstriction) → ↑ PR ⇒ ↑ BP

2. **chemoreceptors** - central (medulla) & peripheral (carotid/aortic bodies)
   - stimulated by ↑ CO_2, ↑ H^+ (↓ pH) & ↓ O_2

   with ↑CO_2, ↑H^+, ↓O_2 → ↑ SNS, ↓ ParaSNS → ↑HR, ↑SV, ↑ vasocon ⇒ ↑ BP
   with ↓CO_2, ↓H^+, ↑O_2 → ↓ SNS, ↑ ParaSNS → ↓ HR, vasodilation ⇒ ↓ BP

   (↓ BP ⇒ ↓ blood delivery; ↑ BP ⇒ ↑ blood (O_2) delivery)

3. **CNS Ischemic Response -** (ischemia = ↓ blood flow)
   - used in emergency situations when BP < 50 mmHg

   ↓ flow → ↑ vasomotor center activity → vasoconstriction ⇒ ↑ BP

** short-term mechanisms lose their ability to regulate BP after hrs/days of deviant BP **

Long-Term Regulation of BP - slow-acting processes

Hormonal control - control persists for periods of min → hours → days

1. adrenal medullla stimulation by SNS (stress, excitement, emergency)

   → ↑ epinephrine release → ↑ HR, ↑ SV, vasoconstriction ⇒ ↑ BP

2. renin → angiotensin → aldosterone

   ↓ BP to kidney (or ↑ K^+) → ↑ renin release by kidney (juxtaglomerular apparatus)

   angiotensin → (in the presence of renin) → angiotensin I →
   (in the presence of angiotensin converter enzyme, lungs) → angiotensin II

   angiotensin II - most **powerful vasoconstrictor** produced by body
   - ↑ vasoconstriction of arterioles & veins (↑ blood flow to kidney)
   - ↑ aldosterone release by adrenal cortex (antidiuretic)
   - ↓ urine volume → ↑ blood volume ⇒ ↑ BP

3. Vasopressin (ADH from hypothalamus) -

   with ↓ BP, ↑ osm → ↑ osmoreceptor stimulation ⇒ ↑ ADH

   ADH → ↑ H_2O reabsorption/vasoconstriction ⇒ ↑ BP

4. ANF (atrial natriuretic factor) - produced by right atrium

   ↑ atrial BP → ↑ ANF → ↑ urine production → ↓ blood volume ⇒ ↓ BP

5. Fluid-Shift mechanism -

   - movement of fluid from interstitial spaces to maintain blood volume

   ↓ BP → move fluid out of interstitial spaces & into capillaries
   → ↑ blood volume ⇒ ↑ BP

   ↑ BP → move fluid out of capillaries/into interstitial spaces
   → ↓ blood volume ⇒ ↓ BP

6. Stress-Relaxation response - adjustment in smooth muscle tone

   with ↑ BP → relax smooth muscle → dilation ⇒ ↓ BP

   with ↓ BP → constriction ⇒ ↑ BP

** Kidney plays an important role in the long-term regulation of BP by regulating blood volume with ↑ or ↓ urine production

Chapter 22 - The Lymphatic System and Immunity

lymph - circulating fluid of lymphatic vessels; ≈ plasma composition

lymphatic system - one-way, "open-ended" system, carries fluid **away from tissues & towards blood veins**
– return system only (no supply)

Functions:
1. **fluid return** - returns fluid that has leaked out of capillaries; **prevents edema**; maintains fluid balance
2. **fat absorption** - occurs through the **lacteals** (lymphatics of small intestine)
3. **defense** - filters lymph through nodes & spleen (filters blood)
   – cells within lymphatic system destroy foreign organisms

lymph capillaries - "blind-end", permeable, fluid enters but doesn't flow out

lymphatic tissue = reticular connective tissue

lymph nodules - small clusters of lymphatic tissue

    Peyer's pouches - located within small intestine

    tonsils - **filters air**, remove foreign materials; located within oro-/nasopharynx
– **pharyngeal (adenoids** when enlarged) - within nasopharynx
– **palatine** - located on either side of oral cavity
– **lingual** - associated with the tongue

lymph nodes - have medulla/cortex regions; **filters lymph** (cervical, inguinal, axillary)

    germinal centers - stimulate lymphocyte development

spleen - **filters blood**
    white pulp - lymphocyte production
    red pulp - phagocytes remove foreign material/worn-out RBCs

thymus - site of **T cell production**
– T cells migrate to other lymphatic tissues where they proliferate
– undergoes **involution** (decreases in size with age)

Immunity - the ability to resist microorganisms & other foreign materials

I. non-specific immunity - "innate" - each encounter gives same response
– born with this ability

II. specific immunity - "adaptive" - **memory** reduces response time
— developed by exposure to foreign material

I. Nonspecific/Innate Immunity - doesn't matter what the foreign material is, the response will be the same

what's involved: A.) mechanical mechanisms that prevent entry
B.) chemical substances
C.) phagocytic cells
D.) inflammatory response

A. mechanical mechanisms -
barriers - skin & mucous membranes
physical processes - **washing** - tears & urine
blowing - cough, sneeze
trap & sweep - involves mucus & ciliated cells

B. chemicals - kill/prevent entry; promote inflammation

complement - 20 plasma proteins (globulins)
— works in a "cascade" once activated (classical and alternative pathways)
— promotes inflammation & phagocytosis; attracts phagocytic cells
— some components directly lyse cells

interferon - protein produced by cells infected with a virus
— stimulates surrounding cells to produce antiviral proteins
— prevents spread of viral agent

surface chemicals -
lysozyme - found in tears & saliva
acid - found in sebum & HCl in stomach
mucus - traps microorganisms

histamines - released from basophils, platelets and mast cells
— ↑ vasodilation, ↑ capillary permeability

kinins - vasodilators, attract neutrophils

PGs (prostaglandins) - cause vasodilation, promote inflammation

pyrogens - cause fever

leukotrienes - promote inflammation

C. **phagocytic cells** -
   neutrophils - phagocytic WBCs, first at site of infection, contain lysosomes
   – **pus** = dead neutrophils + debris

   macrophages - actively phagocytic monocytes
   – ingest more & larger particles than neutrophils
   – seen in late stages of /chronic infections
   – produce interferon, complement & PGs

   basophils - motile WBCs   **mast cells** - nonmotile CT cells
   – activated by complement, release histamine & leukotrienes
   – promote inflammation

   eosinophils - produce antihistamines
   – moderates allergies/inflammatory reactions; kill some parasites outright

   NK (natural killer) cells - type of lymphocyte
   – kill tumor & virus-infected cells

D. **inflammatory response** - initiated by the release/activation of chemicals

   vasodilation - increased blood flow to affected area

   chemotactic attraction of phagocytes
   – phagocytes leave blood & enter tissues by diapedesis

   ↑ **vascular permeability** - complement & fibrinogen enter tissue from bloodstream

   local - confined to specific area - red, hot, swelling, pain, loss of function
   systemic - local symptoms + pyrogens, ↑ [neurophils], ↑ vascular permeability

II. Specific Immunity -
– body recognizes, responds to & remembers pathogen/toxin from previous encounter
– develop memory from exposure → faster 2nd response

Antigens (Ag) - large molecules, stimulate immune response (MW 10,000+)
   foreign - not produced by body
      self - produced by body, important in recognition of self

autoimmune diseases - self Ags stimulate immune response

haptens - smaller molecules, must combine with larger molecules to initiate immune response (e.g., Penicillin)

Types of Specific/Adaptive Immunity

1. **antibody-mediated (humoral) immunity** - due to Abs produced by B lymphocytes
   - most effective against extracellular organisms
   - provide immediate sensitivity
2. **cell-mediated immunity** - involves T lymphocytes
   - most effective against intracellular organisms & delayed hypersensitivity
   - involves **effector T cells** - cytotoxic & delayed hypersensitivity T cells
     & **regulatory T cells** - helper & suppressor T cells (limit/control response)

1. antibody-mediated immunity

lymphocytes - derived from stem cells

clones - group of identical lymphocytes that respond to same Ag

lymphocyte activation - recognize Ag; Ag binds to Ag-receptor on lymphocyte
 – causes ↑ # lymphocytes specific to destroy this Ag (clone)

epitopes = antigenic determinants - region of Ag that activates a lymphocyte

MHC - **m**ajor **h**istocompatability **c**omplex - Ag glycoproteins on cell surface that activate lymphocytes

costimulation - sometimes a 2nd signal is required for activation, after MHC Ags produced
   - involve lymphokines - proteins produced by lymphocytes/regulate activity

tolerance - suppression of immune system response to Ag

Antibodies (Ab) - gamma globulins/immunoglobulins (Ig)
 – produced in response to specific Ag

- 4 polypeptide chains (2L, 2H)
- contain regions for complement binding & macrophage attachment within constant region

variable region - combines with antigenic determinants (epitopes)
 – confers specificity - different for each Ag

IgG - activates complement; opsonization (↑ phagocytosis); crosses placenta; responsible for Rh reaction

IgM - first Ab produced with exposure; activates complement; ABO blood groups

IgA - in breast milk; protects body surfaces (in tears, saliva, mucus)

IgE - binds to mast cells/basophils; ↑ inflammation

IgD - functions as Ag-binding receptors on B cells

Ab **complexes** with Ag - marks Ag for destruction (Table 22.5, Fig. 22.19)

1- interferes with or **inactivates** Ag
2- binds Ag together = **clumping/agglutination** (blood typing, pregnancy tests)
3- **opsonization** - produce opsonins - substances that ↑ phagocytosis by macrophages
4- **activates complement cascade** - inflammation, chemotaxis, lysis
5- initiates **inflammatory response**

B cells ingest Ag, **display** Ag-determinants on CM

Helper T cells bind to Ag-determinants, secrete interleukin (IL-2)

IL-2 stimulates B cells to divide, produce more cells that can produce Ab specific to Ag

primary response - 1st exposure, results in production of **memory** & **plasma** (Ab-producing) cells
    - takes 3-14 days to produce significant amounts of Ab
    - see symptoms & tissue damage

secondary response - memory cells quickly for plasma cells & [Ab] rises quickly
    – response takes hours → few days; no disease symptoms
    – the individual is said to be "immune"

monoclonal Abs - specific production of large amounts of a single Ab
    – produced by **hybridomas** (fused activated T cells + tumor cells)
    – applications - pregnancy tests, disease screening

2. Cell-Mediated Immunity - effective against intracellular organisms

- T cells and macrophages work together
- Ag activates effector T cell → production of memory T cells
- cytotoxic T cells lyse virus-infected cells, tumor cells, transplants
- cytotoxic T cells produce lymphokines → produce phagocytosis, inflammation
- delayed hypersensitivity T cells produce lymphokines → produce allergic reactions

Lymphokines - proteins produced by T cells, activate other immune system components
1- interferon
2- lymphotoxins
3- macrophage-activating factor
4- chemotactic factors
5- interleukin-2

Acquired Immunity - 4 ways to acquire specific immunity

immunization - artificial exposure, deliberate introduction of Ab or Ag

active immunity - individual produces their own Abs

passive immunity - Abs from outside source given to nonimmune person

1- **active natural immunity** - natural exposure to Ag stimulates your system to produce Ab, provides long-lasting protection
2- **active artifical immunity** - deliberate exposure to Ag (e.g., vaccination with Ag)
3- **passive natural immunity** - transfer of Ab across placenta or with milk
4- **passive artificial immunity** - immediate but temporary immunity
    - transfer of Abs from immune → nonimmune individual

Chapter 23 - The Respiratory System

Respiration - O_2/CO_2 exchange; regulates pH

1. **ventilation** - inspiration & expiration
2. **external respiration** - within alveoli

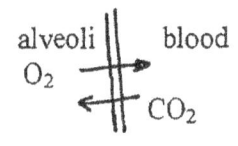

```
alveoli  |  blood
  O₂  →
       ←  CO₂
─────────────────
blood    |  tissues
  O₂  →
       ←  CO₂
```

3. **O_2/CO_2 transport in the blood**
4. **internal respiration** - gas exchange in tissues

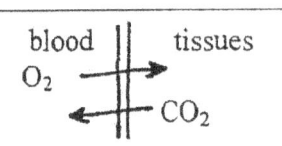

nasal cavity → naso-, oro-, laryngopharynx → larynx → trachea → bronchi (1°,2°,3°)
(nasal conchae,　　　　　　　　　　　　　　(vocal　　　　　　　　↓
 paranasal sinuses)　　　　　　　　　　　　 cords)　　　　　　bronchioles
　　　　　　　　　　　　　　　　　　　　　　　　　　　　　　　　↓
　　　　　　　　　　　　　　　　　　　　　　　　　　　　　　 alveoli

mucus - trap particulates in airstream
cilia - beat, produce directed movement
- within nasal cavity - beat downward　⎫
- within trachea - beat upward　　　　 ⎬ bring to esophagus & swallow
- sneeze (CN V) & cough (CN X) - aid in this process

- air must be warmed & humidified on its way down to the lungs
- nasal mucosa is well-vascularized & performs this function

- each lung (R- 3 lobes; L- 2 lobes) is surrounded by a pleural cavity
  pleural fluid - lubricant - ↓ friction as membranes slide past one another
  - holds visceral/parietal pleura together - ∴ lungs stretch when thorax expands
  　parietal pleura - lines cavity　　　　**visceral pleura** - covers lung surface

trachea - lined with pseudostratified ciliated columnar epithelium (mucociliary escalator)
- reinforced by C-shaped cartilage rings (open except when swallowing)

bronchioles - lined with ciliated epithelium that becomes simple squamous epithelium
- no cartilage support, smooth muscle present (constricts during asthma attack)

alveoli - simple squamous epithelium, performs diffusion function
- **secretory cells** - produce **surfactants**
- **dust cells** - macrophages, defensive function

diaphragm - 1° muscle of ventilation
- when contracted, pulled downward (flattens) → ↑ volume/↓ pressure of thorax
  → **inspiration**
- when relaxed, returns to dome-shaped position → ↓ volume/↑ pressure of thorax
  → **expiration**

Ventilation - requires a pressure gradient
- air flows from high P → low P area
- compare intrapulmonary P vs. atmospheric P
- **Boyle's Law** - P & V are inversely proportional to one another

at end of respiration cycle $P_{intrapulmonary} = P_{atm}$ ∴ no air movement

during inspiration $P_{intrapulmonary} < P_{atm}$ ∴ air moves into lungs
↑V_{lungs} → ↓P_{lungs}

during expiration $P_{intrapulmonary} > P_{atm}$ ∴ air moves out of lungs

lungs contain elastic tissue → stretch/recoil with changing thoracic volume

inspiratory muscles contract → ↑ V_{thorax} → ↑ V_{lungs} → ↓ $P_{intrapulmonary}$
                                            (stretch)

diaphragm relaxes → ↓ V_{thorax} → ↓ V_{lungs} → ↑ $P_{intrapulmonary}$
                                   (recoil)

Factors preventing lung collapse:

1- **surfactants** - reduce surface tension within alveoli
        - without surfactants → respiratory distress syndrome (RDS)
2- **negative intrapleural P** - P within pleural cavity ~ 2mmHg below P_{atm}
3- **residual volume** - air that remains in the lungs after expiration

    pneumothorax - air introduced into the pleural cavity → lungs collapse

Pulmonary volumes - measured with spirometry

    TV (tidal volume) = 500 ml   normal ventilation volume
    IRV = 3000 ml   forced inspiration above TV
    ERV = 1100 ml   forced expiration beyond TV
    RV = 1200 ml   air remaining in lungs after forced expiration

Inspiratory capacity (IC) = TV + IRV = 3500 ml

Functional residual capacity = ERV + RV = 2300 ml
  = air left in lungs after normal expiration

Vital capacity (VC) = IRV + ERV + TV = 4600 ml
  = maximum ventilation volume

Total lung capacity (TLC) = IRV + ERV + TV + RV = 5800 ml
  = total amount of air lungs can hold

Minute respiratory volume (MRV) = TV x respiration rate (# breaths/min)
  = 500 ml/breath x 12 breaths/min = 6 L/min

Anatomical/physiological dead space = 150 ml
  = air not available for external respiration

Alveolar ventilation rate = respiration rate x (TV - dead space)
  = 12 breaths/min x (500 ml - 150 ml) = 4.2 L/min

+ review muscles of respiration
+ quiet vs. forced respiration - inspiration & forced expiration = active processes, ATP
  – quiet expiration = passive process, no ATP

Physical Principles of Gas Exchange (Fig. 23.16)

P_{atm} = 760 mmHg O_2 ~ 21% of air mixture, P_{O2} = 160 mmHg

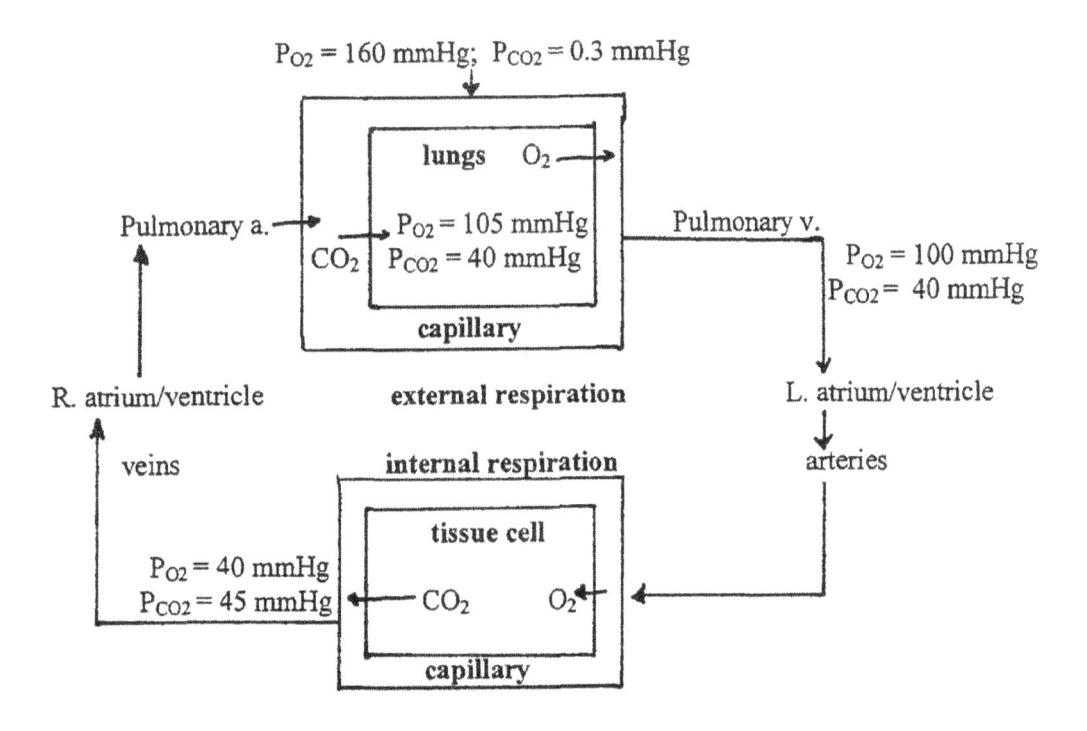

P_{O2} = 160 mmHg; P_{CO2} = 0.3 mmHg

lungs O_2

Pulmonary a. → P_{O2} = 105 mmHg   Pulmonary v.
         CO_2 P_{CO2} = 40 mmHg P_{O2} = 100 mmHg
                                                  P_{CO2} = 40 mmHg
         capillary

R. atrium/ventricle   **external respiration**   L. atrium/ventricle

veins   **internal respiration**   arteries

                tissue cell

P_{O2} = 40 mmHg
P_{CO2} = 45 mmHg ← CO_2 O_2 ←

                capillary

compliance - measure of **expansibility** of lungs and thorax
— emphysema/COPD, RDS, fibrosis, etc. affect compliance

respiratory membrane - thin, large surface area; facilitates gas exchange
components: + surfactant/fluid layer
+ alveolar simple squamous epithelium
+ basement membrane
+ thin interstitial space
+ blood capillary basement membrane
+ blood capillary endothelium (simple squamous epithelium)
+ RBC membrane

factors that influence the **rate of gas exchange** across the respiratory membrane:
+ **thickness** - ↑ thickness → ↓ rate of diffusion
+ **surface area** - ↓ surface area → ↓ rate of diffusion
+ **partial pressure difference** - as ↑ $P_1 - P_2$ → ↑ rate of diffusion
+ **diffusion coefficient** - depends upon the size/solubility of O_2/CO_2 in H_2O
— CO_2 is 24x more soluble than O_2 in H_2O

O_2/CO_2 transport in blood

O_2 - 97% bound to hemoglobin; 3% dissolved in plasma

CO_2 - 8% dissolved in plasma
— 20% bound to hemoglobin; carbaminohemoglobin
— 72% in plasma as HCO_3^-

✳ $CO_2 + H_2O$ <u>carbonic anhydrase</u> → H_2CO_3 ⇌ $H^+ + HCO_3^-$ (within RBC)

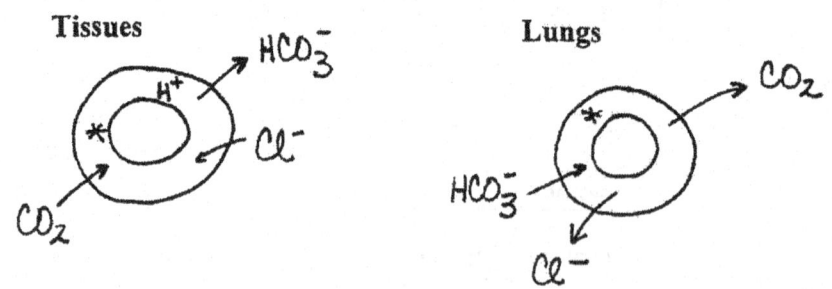

Cl⁻ shift - movement of Cl⁻ into (tissue)/out of (lungs) RBC in exchange for HCO_3^-

H^+ binds to Hb → HHb, reduced Hb; prevents O_2 binding, helps prevent drop in pH

O₂-Hb dissociation curve – describes the percent of Hb saturation w/O₂ at any pO₂

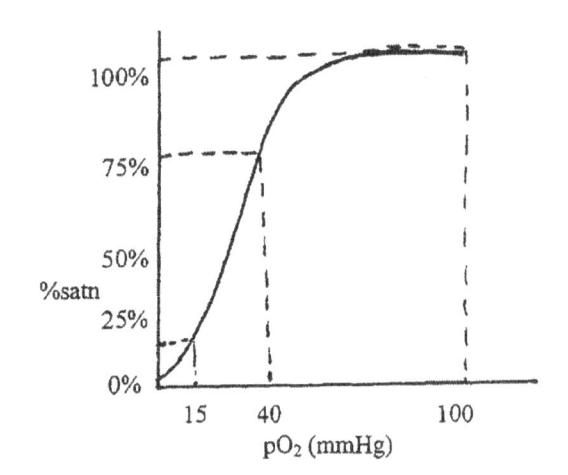

sigmoidal curve = cooperativity

dissociation curve influenced by: pH, CO₂ & T

↓pH, ↑CO₂, ↑T, ↑BPG; shift to right (↓ % saturation)

↑pH, ↓CO₂, ↓T, ↓BPG; shift to left (↑ % saturation)

pO₂(lungs) = 100 mmHg;   pO₂(tissues at rest) = 40 mmHg;   pO₂(exercise) = 15 mmHg

in systemic arterial blood, ~100% saturation (each Hb has 4 O₂)
in tissues (at rest), ~75% saturation (each Hb loses 1 O₂ during internal respiration)
in tissues (exercise), ~25% saturation (3 O₂ → tissues; 1 O₂ remains bound to Hb)

BPG = 2,3 bisphosphoglycerate, produced by RBC, ↓ O₂ binding affinity of Hb

CONTROL OF RESPIRATION

1- **Nervous control** -
   a) **medulla** - contains respiratory centers
      <u>inspiratory neurons</u> - I neurons
         – stimulate inspiratory muscles to contract, establish rhythm

      <u>expiratory neurons</u> - E neurons
         – no active during normal breathing (quiet expiration is passive)
         – only necessary in forced expiration (additional muscles must contract)

   b) **pons** - reinforce breathing rhythm with pontile centers
      <u>pneumotaxic center</u> - inhibits I neurons, inhibits apneustic center,
                     promotes exhalation

      <u>apneustic center</u> - continuously stimulates I neurons

   c) **stretch receptors** - in lungs; inflated lungs send inhibitory signal to I neurons
                     causing exhalation

<u>Hering-Breuer reflex</u> - depends on stretch receptors; prevents overinflation of lungs (prevents damage to elastic tissue)

   d) **conscious control** of ventilation - willfulness, pain, touch, T alter ventilation

2- **Chemical control -**
   a) **CO_2/pH** - CO_2 is primary regulator of respiration

      ↑ CO_2 signal picked up by CNS chemoreceptors within medulla
      ↑ CO_2 (hypercapnia) leads to ↓ pH (↑ H^+) → ↑ ventilation → blow off excess CO_2

   b) **O_2** - has less influence - must experience ~Δ50% before signals received
     ↓ O_2 (hypoxia) picked up by peripheral chemoreceptors → ↑ ventilation

Chapter 24 - The Digestive System

GI tract = "tube within a tube"

Functions: **ingestion** - intake
propulsion - movement of food (24 - 36 hrs)
digestion - mechanical & chemical breakdown
absorption - from lumen of GI tract → bloodstream
excretion - elimination of wastes

4 lays (tunics) of GI tract -

1- **mucosa** - innermost layer, in contact with food in lumen
– contains 3 layers: **epithelium** - stratified squamous (mouth → esophagus);
simple columnar (stomach →)
lamina propra - CT
muscularis mucosa - smooth muscle

2- **submucosa** - thick CT with nerves, blood vessels & glands
– **submucosa plexus** - nerves with parasympathetic cell bodies

3- **muscularis** - skeletal muscle (oral cavity → top 1/3 esophagus)
smooth muscle (bottom 2/3 esophagus →)
– inner layer (circular smooth muscle), outer layer (longitudinal smooth muscle)
– stomach has 3rd layer with oblique fiber direction
– **myenteric plexus** - nerves found between muscularis layers
– **intramural plexus** - submucosal + myenteric plexus; controls movement of materials & secretions

4- **serosa (adventitia)** = visceral peritoneum, outermost layer

Functions associated with the oral cavity:

mastication - mechanical process, increases surface are of food
– conversion of food → bolus
– involves muscles - masseter, pterygoid, temporalis

3 salivary glands - parotid, submandibular, sublingual
– exocrine glands, produce 1.5 liters/day

saliva = **mucus** - binds & lubricates material
salivary amylase - enzyme that begins carbohydrate digestion
lysozyme - antibacterial
Ig A - an antibody

deglutition = swallowing; voluntary (mouth) vs. involuntary (pharynx, 2 sec & esophagus, 8 sec)
— epiglottis covers opening into larynx during swallowing

deglutition apnea - reflex preventing breathing during swallowing

Esophagus - muscular tube from pharynx → stomach
— sphincters at top & bottom
— **peristalsis** - waves of muscular contraction
— mucus & gravity assist fast movement of bolus

Stomach - primary storage & mixing chamber
— minimal digestion & absorption takes place
— bolus → chyme conversion
— begins protein digestion; no CHO/lipid digestion; pH2

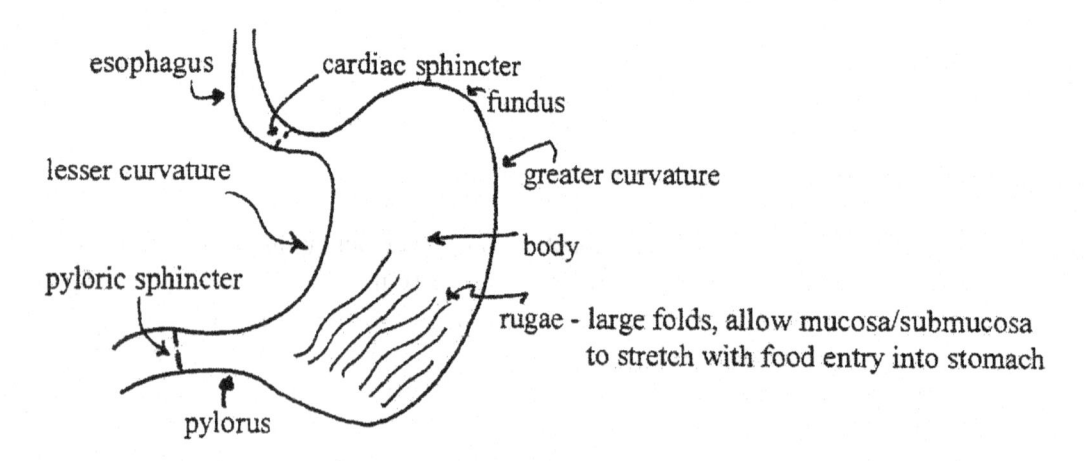

[Diagram of stomach with labels: esophagus, cardiac sphincter, fundus, lesser curvature, greater curvature, body, pyloric sphincter, pylorus, rugae - large folds, allow mucosa/submucosa to stretch with food entry into stomach]

Specialized stomach cells and their secretions:

1- **mucous cells** - secrete an alkaline mucus layer; lubricates & protects

2- **parietal cells** -
   secrete **intrinsic factor** - glycoprotein binds with vitamin B_{12}, aids in its absorption within the small intestine

   b) produce **HCl** - low pH of stomach contents kills bacteria & denatures proteins

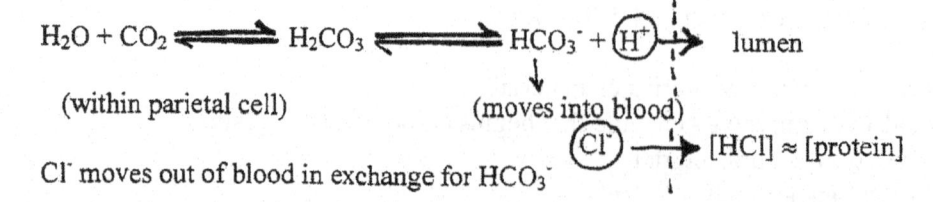

$$H_2O + CO_2 \rightleftharpoons H_2CO_3 \rightleftharpoons HCO_3^- + H^+ \rightarrow \text{lumen}$$

(within parietal cell)   (moves into blood)
                                    $Cl^- \rightarrow$ [HCl] ≈ [protein]

Cl^- moves out of blood in exchange for HCO_3^-

3- **chief cells** - secretes inactive (zymogenic) pepsinogen

$$\text{pepsinogen} \xrightarrow{\text{HCl}} \text{pepsin (proteolytic enzyme)}$$

4- **G cells** - produce **gastrin**; a hormone that stimulates HCl secretion & gastric motility/emptying

stomach secretions can be regulated -
    cephalic - sensory stimuli → Ach release, ↑ gastric secretion
    gastric - food in stomach → ↑ gastric secretions
    intestinal - controls entrance of chyme → small intestine

mixing action of stomach - a function of the 3 layers of muscularis
- mixing waves & strong peristaltic waves are involved
- **pyloric pump** - functions in gastric emptying

Small Intestine - primary site of chemical digestion & absorption
    stomach → duodenum → jejunum → ileum → large intestine

    plicae circulares - permanent folds/ridges in lining of small intestine
- chyme spirals down these folds
- increase surface area within small intestine

    villi - fingerlike projections of mucosa; contain blood vessels & lacteals

    lacteals - lymphatic drainage vessel, run up within villi; remove chylomicrons

chylomicrons - droplets containing digested fat

microvilli - "brush border"; folding of epithelial CM of villi, ↑ absorptive surface area

brush-border enzymes - found within epithelial cells of SI, released into lumen when cells slough off

goblet cells - produce mucus

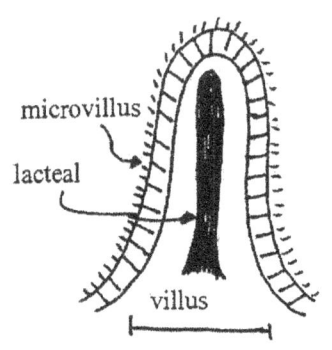

movement within the small intestine -
    segmentation - alternate contraction/relaxation cycles & some peristalsis

secretin & cholecystokinin (CCK) - hormones produced by the small intestines
- ↑ pancreatic secretions, ↑ bile secretion, ↑ mucus secretion
- CCK causes contraction of gall bladder & relaxation of ampulla of Vater
- release of "brush border" enzymes

- duodenum receives hepatic (bile) & pancreatic secretions through ducts that open @ ampulla of Vater (hepatopancreatic ampulla)

Liver - multiple functions performed:
1- bile production
2- remove sugar from blood; stores excess glucose as glycogen
3- storage of fats, vitamins and minerals
4- interconversion of nutrients (e.g., aa → glucose; activation of vitamin D)
5- detoxification - e.g., alcohol, ammonia
6- phagocytosis - **kupffer cells** - liver macrophages
7- manufacture proteins - albumin, clotting factors, fibrinogen, heparin

Bile - mixture of bile salts, pigments, cholesterol & bilirubin; contains NO enzymes
– **emulsifying agent** - breaks up fats into smaller droplets, increases surface area
– reabsorbed/recycled within small intestines - enterohepatic circulation (~80%)

Gallbladder - stores and concentrates bile
  gallstones - cholesterol, etc.; precipitate within gallbladder and ducts

  bile drainage: R/L hepatic ducts → common hepatic duct
                                              ↓
                                         gallbladder
  ampulla of Vater ← common bile duct ←   ↓
                                         cystic duct

Pancreas - exocrine portion of gland contains acinar & intercalated duct cells

acinar cells

duct cells   **acinus** = secretory unit, produces pancreatic juice
                       with aqueous and enzymatic components

pancreatic duct

aqueous component - produced by intercalated duct cells
  – contains HCO_3^- to neutralize acidic chyme as it enters small intestine
  – increases pH within small intestines to ~ pH 7-8, the appropriate pH for pancreatic enzyme action

enzyme component - produced by acinar cells
– contains enzymes to digest CHO, lipids & proteins

1. **proteolytic enzymes** - secreted in inactive (zymogenic) form to protect pancreas from autodigestion

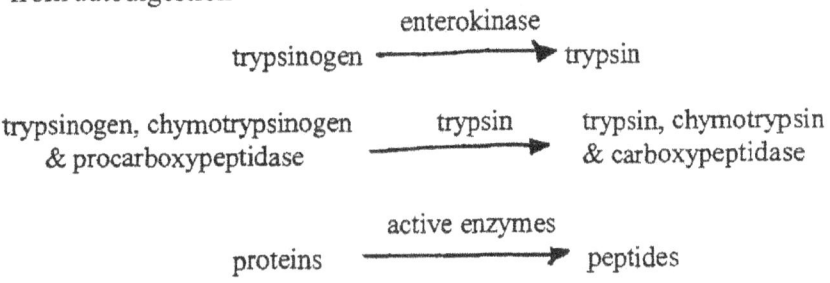

$$\text{trypsinogen} \xrightarrow{\text{enterokinase}} \text{trypsin}$$

$$\text{trypsinogen, chymotrypsinogen \& procarboxypeptidase} \xrightarrow{\text{trypsin}} \text{trypsin, chymotrypsin \& carboxypeptidase}$$

$$\text{proteins} \xrightarrow{\text{active enzymes}} \text{peptides}$$

2. **pancreatic amylase** - digests carbohydrates

3. **pancreatic lipase** - digests lipids → FA, glycerol, cholesterol

4. **DNAase, RNAase** - digest nucleic acids

parasympathetic - ↑ pancreatic secretion;   sympathetic - ↓ pancreatic secretion

Large Intestine = Colon
– contains haustra (pouches) and taeniae coli
– cecum → ascending colon → transverse colon → descending colon → sigmoid colon → rectum
– slower movement of material; chyme → feces
– site of mucus secretion; water/salt absorption
– HCO_3^- secreted to neutralize acidic waste products of bacteria
– vitamin K produced by colon bacteria (vit K important for blood clotting)
– colon bacteria produce organic solids & gases (CH_4, H_2S) = flatus
– feces = water, undigested materials, microorganisms, etc.
– primary propulsion in large intestine = mass movement

mass movement - strong peristaltic contractions (3-4x day)

■ in colon, materials spend most time in ascending segments (moving against gravity)

Digestion - know appropriate enzymes, substrates, end-products & locations

Absorption - know locations & transport mechanisms used

DIGESTION:

Mouth carbohydrates —salivary amylase→ carbohydrates

Stomach proteins —HCl, pepsin→ dipeptides, polypeptides

Duodenum CHO —pancreatic amylase→ disaccharides —lactase, maltase, sucrase (brush border enz)→ simple sugars

polypeptides & dipeptides —trypsin, chymotrypsin, carboxypeptidase, aminopeptidase (brush border enz)→ dipeptides & aa

fats —bile→ emulsified fats —pancreatic lipase→ FA + glycerol

ABSORPTION

simple sugars + amino acids → absorption into intestinal capillaries by facilitated diffusion and cotransport

FA + glycerol —bile salts→ micelles → chylomicrons → lacteals

micelles - triacylglycerol vesicles

chylomicrons - protein-coated triacylglycerol droplets

Chapter 25 – Nutrition, Metabolism and Temperature Regulation

Nutrients - used to produce energy, provide building blocks, etc.

Metabolism - chemical changes that occur within the body
  Anabolism - synthetic rxns; energy-requiring
  Catabolism - breakdown rxns; energy-releasing

Vitamins - essential for normal metabolism, function as coenzymes
  fat-soluble - vit A,D,E,K; cross CM can accumulate in toxic amounts
  water-soluble - vit C,B complex; ingested excess quickly excreted from the body

Minerals - Ca^{2+} - important in bone, muscle contraction, blood clotting, nerve/heart fcn
  Fe - part of Hb
  Na^+/K^+ - osmotic balance; irritability of CM

Digestion - mechanical/chemical breakdown of food into simple nutrients

Absorption - nutrients pass through all 4 tunics of gut to reach hepatic portal vessel or lacteals

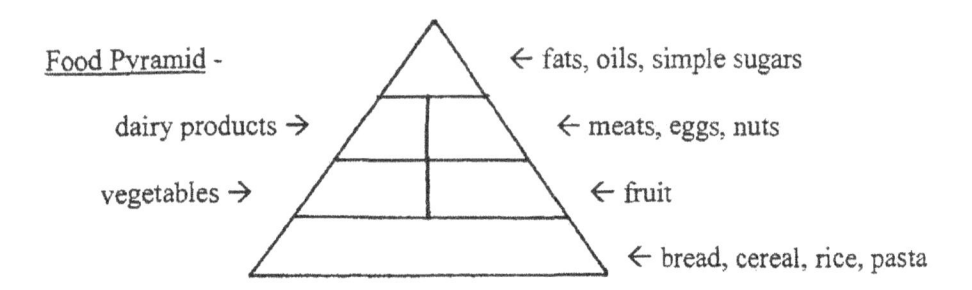

Food Pyramid -
  ← fats, oils, simple sugars
  dairy products → ← meats, eggs, nuts
  vegetables → ← fruit
  ← bread, cereal, rice, pasta

essential nutrients - must be ingested because the body cannot make them

carbohydrates - most common CHOs in diet are glucose & fructose, starch, sucrose
  & cellulose (nondigestible, "bulk", aids in GI motility)

  - recommended amts 125-175 g/day; complex rather than simple sugars suggested
  - CHO digestion yields 4 cal/gm

lipids - ~ 95% in diet are triacylglycerols; saturated and unsaturated fats
  ~ 5% are cholesterol (animal product) & phospholipids (lecithin in egg yolks)

  - recommended amt < 30% of total caloric intake; saturated fats should be no more than 10% of total fat intake
  - keep cholesterol ≤ 300 mg/day
  - cholesterol & saturated fats promote coronary artery disease
  - lipid digestion yields 9 cal/gm

essential FA - must be ingested, used to synthesize PGs
  eg. linoleic acid, EPA (eicosapentaenoic acid) - found in plant oils

proteins - essential aa - 9 aa that cannot be synthesized by the body
  nonessential aa - remaining 11 aa, can be made by the body

complete protein - contains adequate amounts of the 9 essential aa;
  primarily animal products - milk, meats, egg, cheese

incomplete protein - lack 1 or more of the essential aa;
  primarily plant products - potential problem for vegetarians

recommended amt of protein in diet = 0.8g/kg body wt, ~ 12% of total calories
protein digestion yields 4 cal/g

N balance - amt N ingested ≈ amt N excreted
  children should have positive N balance - N needed for growth, protein synthesis

vitamins - needed in small amts
  vit K = blood clotting   vit B_{12}/folic acid - RBC synthesis
  vit D = Ca/P metabolism   vit C - collagen synthesis; protein metabolism
  vit A = rhodopsin

hypervitaminosis - toxic level of vitamins accumulate

Glycolysis of glucose (within cytoplasm)

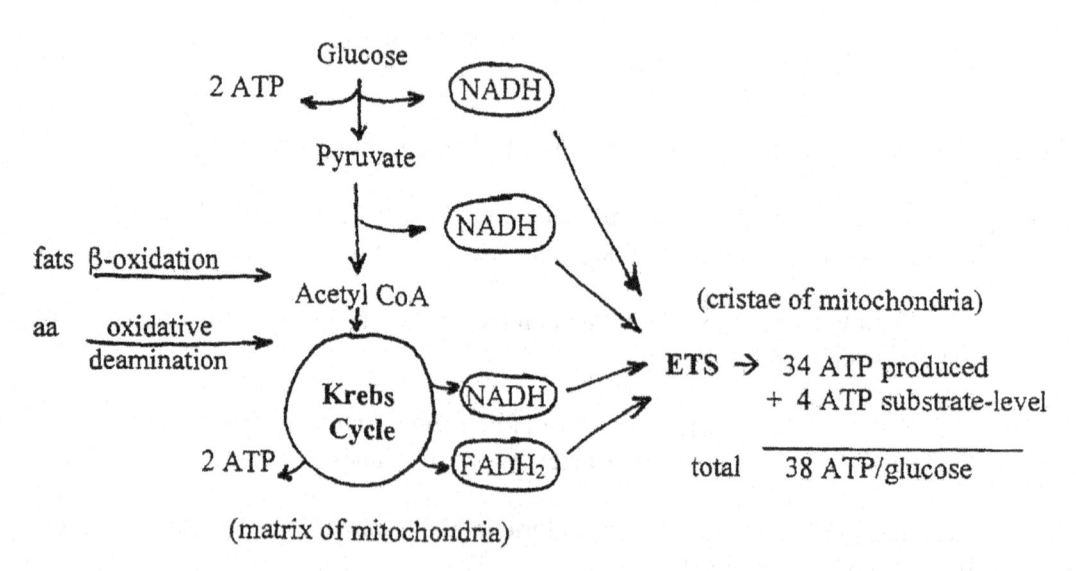

```
                Glucose
         2 ATP ⇄ ↓ → NADH
                Pyruvate
                   ↓ → NADH
fats  β-oxidation
                Acetyl CoA                    (cristae of mitochondria)
aa   oxidative
     deamination         
                   ↓                          ETS →  34 ATP produced
                 Krebs  → NADH                      + 4 ATP substrate-level
                 Cycle                              ─────────────────────
         2 ATP ← ↘ → FADH₂              total    38 ATP/glucose
         
         (matrix of mitochondria)
```

glycolysis occurs w/ or w/o oxygen (w/o oxygen, lactate accumulates)
Krebs cycle and ETS require oxygen

oxidative phosphorylation - uses ETS; oxygen serves as terminal e⁻ acceptor
- large quantities of ATP are produced

interconversion of nutrients - occurs primarily in the liver
    glycogenesis = glucose → glycogen
    lipogenesis = excess aa + glucose → lipids
    glycogenolysis = glycogen → glucose
    gluconeogenesis = aa + glycerol → glucose

2 major metabolic states:
- **absorptive state** - lasts about 4 hrs after a meal; nutrients absorbed through intestines & utilized for energy
- **postabsorptive state** - follows absorptive state; stored molecules used for energy; blood sugar ~ 70-110 mg/100ml

metabolic rate - total amount of energy produced/used per unit time

- energy available in foods, released through metabolism & is measured in **calories**
- 3500 cal = 1 lb of body fat
- amount of calories needed to maintain resting body function:
  males = wt x 12; females = wt x 11
- amount of calories necessary based on activity levels:
  inactive = wt x 10
  moderately active = wt x 15
  very active = wt x 20

Calorie = unit of heat content or energy. The quantity of energy required to raise the temperature of 1 gram of H_2O by 1°C

Body Temperature Regulation

1- Body temperature is a balance between heat gain & heat loss
- heat produced through metabolism
- heat exchanged through radiative cooling, conduction, convection & evaporation

2- Rate of heat exchange depends upon T difference between body & surroundings; greater ΔT → greater rate of heat exchange

3- Body's "thermostat" is located within the hypothalmus; maintains body temperature around a "set point"

Chapter 26 - The Urinary System

Functions - removal of waste products from blood
- controls blood volume and BP
- regulation of ion concentration (Na^+, K^+, HCO_3^-)
- regulation of blood pH
- controls RBC production (erythropoeitin)
- controls vitamin D synthesis (skin, UV light)

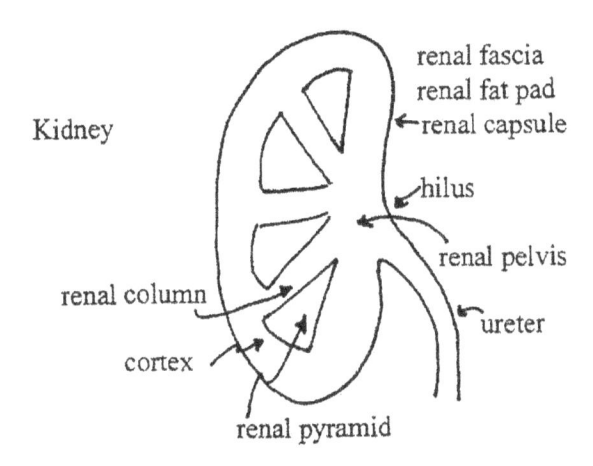

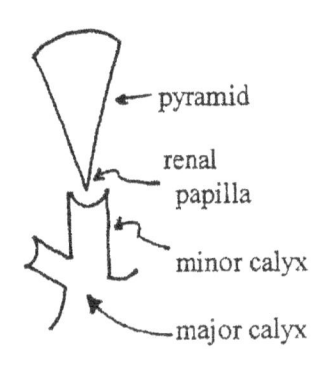

Kidney - renal fascia, renal fat pad, renal capsule, hilus, renal pelvis, ureter, renal column, cortex, renal pyramid

pyramid, renal papilla, minor calyx, major calyx

Nephron - functional unit of kidney
- **cortical** - ~85%, all components within cortex
- **juxtamedullary** - ~15%, Loop of Henle extends into medulla

~ 1.3×10^6 nephrons/kidney
~ 1/3 must function/kidney to ensure survival

nephron = glomerulus + tubular portion

glomerulus = fenestrated capillary bed, site of filtration
tubular portion = site of reabsorption & secretion
= Bowman's capsule → PCT → descending/ascending Loop of Henle
→ DCT → collecting duct

renal corpuscle = glomerulus + Bowman's capsule

Bowman's capsule - visceral layer has specialized cells = **podocytes**

filtration slits - gaps between podocyte processes surrounding the glomerulus
filtration membrane - capillary epithelium + basement membrane + podocytes

[Diagram of nephron showing glomerulus, Bowman's capsule, PCT, descending limb, Loop of Henle, ascending limb, DCT, and collecting duct]

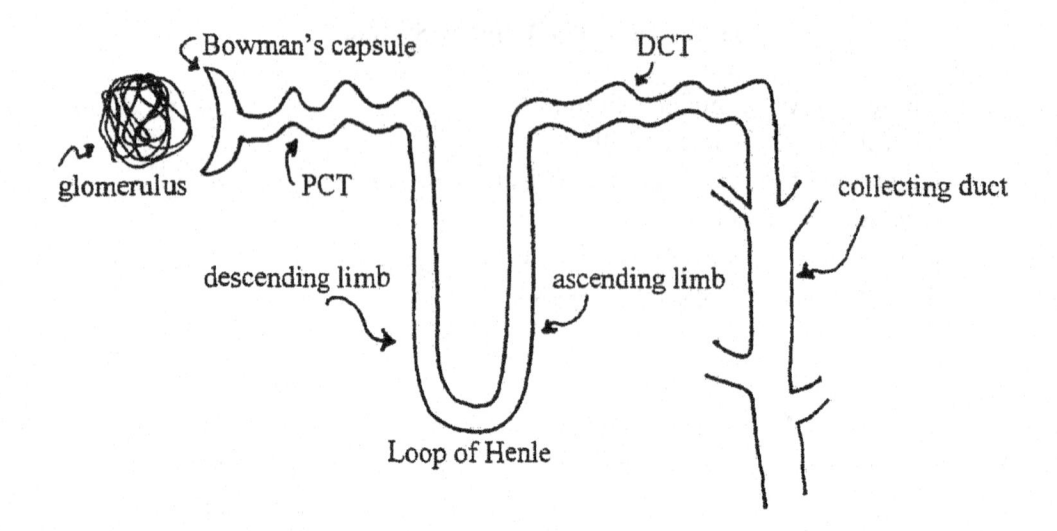

juxtaglomerular apparatus = JG cells + macula densa

JG cells - modified smooth muscle cells of afferent arteriole
– release renin with ↓ BP, results in the restoration of ↑er BP

macula densa - specialized DCT cells

renal a → segmental a → interlobar a → arcuate a → interlobular a → afferent arteriole
↓
glomerulus
↓
efferent arteriole
↓
IVC ← renal v ← interlobar v ← arcuate v ← interlobular v ← peritubular capillaries

glomerulus - site of filtration
peritubular capillaries - site at which reabsorbed materials reenter bloodstream
vasa recta - specialized parts of peritubular capillaries that wrap around Loop of Henle

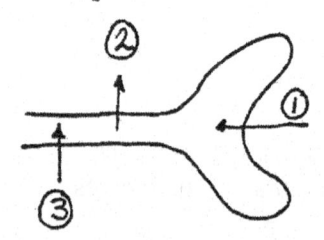

Urine production involves:
1- **filtration** - movement of materials across filtration membrane due to pressure differences
2- **reabsorption** - movement of materials from filtrate back into bloodstream
3- **secretion** - active transport of additional materials into nephron

glomerular filtration - high glomerular capillary pressure forces fluid out of capillary & into Bowman's capsule creating the filtrate

glomerular filtration rate (GFR) = 125 ml/min = 180 L/day

99% of filtrate is reabsorbed ∴ 1.25 ml/min or 1.8 L/day is excreted as uring

filtrate must cross filtration membrane:
- capillary epithelial cells - fenestrated
- basement membrane
- visceral layer of Bowman's capsule - through podocytes with filtration slits

filtration is **nonselective** - anything < 7 nm diameter or < 40,000 MW passes through
 – albumins ≥ 7 nm, pass through to filtrate in small quantities
 – larger plasma proteins are retained

glomerular filtration pressure (GFP)

GCP = 60 mmHg
CP = 18 mmHg
COP = 32 mmHg

GCP = glomerular capillary P
CP = capsule P
COP = colloidal osmotic P

GFP = GCP (60 mmHg) - CP (18 mmHg) - COP (32 mmHg) = 10 mmHg

GCP ≡ hydrostatic P, forces fluid out of glomerulus
CP ≡ caused by fluid already present in Bowman's capsule
COP ≡ plasma proteins remaining within glomerulus draw fluid to them

tubular reabsorption - filtrate leaves renal tubules and reenters bloodstream via peritubular capillaries
 – occurs via active and passive processes

PCT - proteins leave by endocytosis
 – active transport (or co-transport) removes glucose, aa, Na^+, Cl^-, K^+, Mg^{2+}, Ca^{2+}, P

Loop of Henle (ascending limb) - active transport (or cotransport) - Na^+, Cl^-, K^+

DCT - active/co-transport of Na^+, Cl^-, Ca^{2+}

- by the end of the PCT - filtrate volume is reduced by 65%
- Loop of Henle (descending limb is H_2O permeable) - reduces filtrate another 15%
- DCT & CD - H_2O permeability is affected by ADH
    – with ADH another 19% of the filtrate is reabsorbed

∴ urine volume ≅ 1% of filtrate volume

tubular secretion - can be active or passive; movement of materials out of peritubular c capillaries and into renal tubules

PCT - active transport - H^+, NT, toxins, drugs, bile pigments

DCT $<$ active - K^+

passive - H^+, K^+

hyperosmotic urine - humans produce concentrated urine due to ability to maintain high medullary concentration gradient

countercurrent multiplier system - concentrates urine

countercurrent flow = fluid flowing in parallel tubes but in opposite directions
    – materials move from one tube segment to the next
    – egs., loop of Henle and vasa recta

vasa recta - removes excess H_2O/solutes from the interstitial fluid of the medulla without changing the high Osm of fluid within the medulla of the kidney

Loop of Henle - H_2O moves out of descending limb and into the vasa recta
    – in ascending limb, NaCl moves into interstitial fluid to maintain its high Osm

CD - H_2O moves out of CD under the influence of ADH

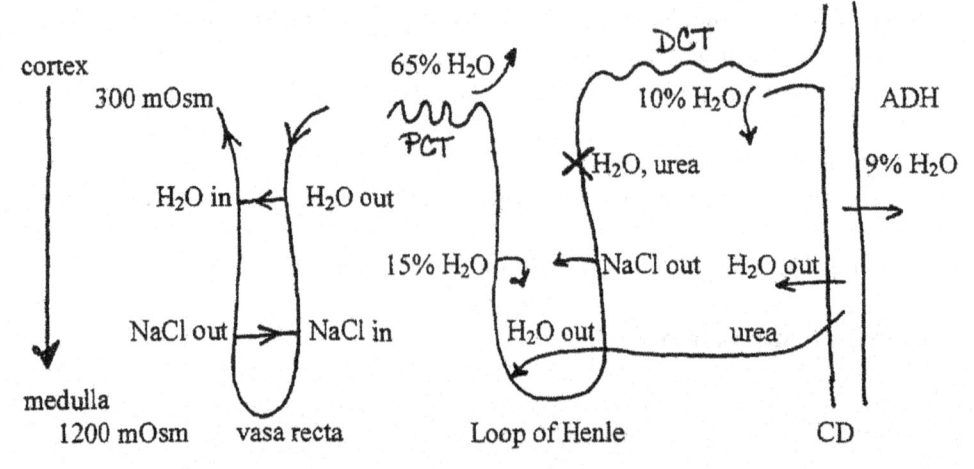

urea - diffuses into descending limb of Loop of Henle from interstitial fluid
  – ascending limb and DCT are impermeable to urea
  – CD is permeable to urea; urea moves out of CD & into interstitial fluid

REGULATION OF URINE PRODUCTION

I. Hormonal control - a total of 4 hormones are involved

1. **aldosterone** (adrenal cortex) - stimulated by ↑ [K^+] & angiotensin II
   → Na^+ reabsorption, H_2O reabsorption/K^+ secretion ⇒ ↑ blood volume → ↑ BP
   ↓ urine volume

2. **renin-angiotensin** - stimulated by ↓ BP
   ↓ BP → renin release by JG cells → angiotensinogen <u>renin</u> angiotensin I
   ACE (angiotensin converting enzyme) ↓
   angiotensin II

   angiotensin II - vasoconstriction (↑ BP), aldosterone release,
   stimulates thirst center in brain

3. **ADH** - produced with ↑ Osm, ↓ blood volume (↓ BP)
   ↑ permeability of DCT/CD to H_2O → ↑ H_2O reabsorption → ↑ blood volume → ↑ BP
   ↓ urine volume

   without ADH - excrete large amounts of dilute urine

4. **ANF** - produced by right atrium with ↑ BP; inhibits ADH secretion
   ↓ ADH → ↑ urine volume, ↓ BP/blood volume

II. Autoregulation - maintenance of stable glomerular filtration rate

   with ↑ BP → afferent arteriole constricts → ↓ blood flow to glomerulus
   with ↓ BP → afferent arteriole dilates → ↑ blood flow through glomerulus

III. Sympathetic innervation - SNS stimulation constricts afferent arteriole
   → ↓ blood flow & ↓ filtrate formation

diuretics - ↑ urine volume ∴ ↓ blood volume & ↓ BP
— many prevent Na^+ reabsorption ∴ Na^+/H_2O lost to urine
— "K^+ sparing" diuretics - prevent excessive K^+ loss

EtOH - ↓ ADH; **caffeine -** ↓ Na^+/Cl^- reabsorption; ↑ GFR

dialysis - doesn't occur naturally; replaces function of nephron (i.e., filtration, reabsorption and secretion)

renal calculi - kidney stones; contain Ca^{2+} salts; deposit within renal pelvis
lipotripsy - stones shattered and passed/aspirated

micturition reflex - elimination of fluid from bladder; stimulated by volumes of ~300 ml

bladder controlled by internal/external urinary sphincters
↙   ↘
smooth muscle   skeletal muscles (conscious control)

stretch receptors of bladder → spinal cord → parasympathetic outflow (relaxes sphincters)
↓
higher brain centers → voiding

(can inhibit relaxation of external sphincter delaying voiding)

Chapter 27 - Water, Electrolyte & Acid-Base Balance

Maintenance of water, electrolyte & acid-base balance involves:
kidneys, lungs, GI tract, and skin
- control is coordinated by nervous & endocrine systems

2 major fluid compartments:

intracellular fluid (ICF) - inside cells, accounts for ~ 40% of total body weight

extracellular fluid (ECF) - outside cell, e.g., plasma, interstitial fluid, etc.
- accounts for ~ 20% of total body weight

composition of ICF/ECF is maintained by CM dynamics:

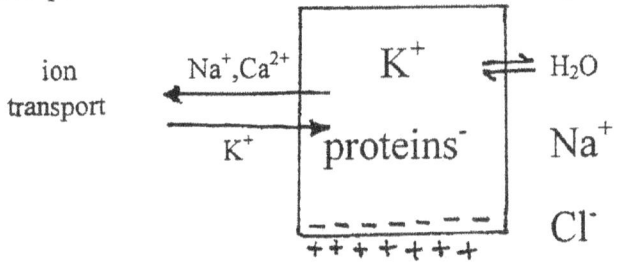

ion transport: Na^+, Ca^{2+} ← ; K^+ → ; K^+ inside ; proteins⁻ ; H_2O ⇌ ; Na^+ ; Cl^-

1- proteins inside cell are too large to move through CM
2- polar substances are transported across CM by carrier molecules (e.g., facilitated diffusion - aa, glucose) & active transports (e.g., ion pumps)
3- charge differential across CM - membrane is polarized
4- water moves by osmosis into/out of cell

for homeostasis to occur: intake = elimination

Regulation of ion concentrations:

Na^+ - primary extracellular cation, responsible for ~90% of osmotic P within ECF

↑ Na^+ → ↑ Osm → H_2O moves into ECF to dilute Na^+ → ↑ blood volume → ↑ BP

↑ BP → ↑ ANP → → ↑ Na^+ excretion/H_2O loss → ↓ BP

↑ Osm, ↓ BP → ↑ ADH → ↑ H_2O reabsorption → ↑ BP, ↓ Osm

↓ BP → renin → → angiotensin II → aldosterone → ↑ Na^+/H_2O reabsorption → ↑ BP

↑ K^+ → aldosterone → ↑ K^+ secretion → ↓ K^+

kidney - primary organ controlling [Na^+] within ECF

hypernatremia - ↑ blood Na^+, seen with hypersecretion of aldosterone

hyponatremia - ↓ blood Na^+, seen with hyposecretion of aldosterone

hyperkalemia - ↑ blood K^+, seen with hyposecretion of aldosterone

hypokalemia - ↓ blood K^+, seen with hypersecretion of aldosterone

Cl^- - primary extracellular anion; attracted to Na^+ in ECF; passively follows Na^+ movt

K^+ - primary intracellular cation
- $[K^+]$ within ECF kept low & maintained within very narrow range
- $[K^+]$ has major influence on RMP
- aldosterone responsible for $[K^+]$

with ↑ $[K^+]_{ECF}$ → depolarization   with ↓ $[K^+]_{ECF}$ → hyperpolarization

Ca^{2+}/(P) - concentrations are controlled by the kidney

↓ Ca^{2+} → ↑ PTH → ↑ Ca^{2+} ↑ Ca^{2+} → calcitonin → ↓ Ca^{2+}

water - ~ 50% H_2O (adult females) < ~ 60% H_2O (adult males)
- due to ↑ [fat] in females & ↑ [protein] in males

Δ H_2O volume affects Osm & BP   ↑ H_2O → ↑ BP, ↓ Osm
                                                                       ↓ H_2O → ↓ BP, ↑ Osm

H_2O intake = H_2O loss   to maintain homeostasis

H_2O loss via: urine, feces, evaporation ⟨ lungs / skin ⟨ insensible perspiration - continual
                                                                                                            sensible perspiration - H_2O/ion loss

H_2O intake via: food + metabolic water (made by the body as a result of chemical rxns)

- kidneys are the primary organs that regulate H_2O volume
- ADH is the primary hormone in regulating H_2O balance

Acid-Base Balance - pH arterial blood = 7.4; pH venous blood = 7.35
- lower pH of venous blood due to ↑ [CO_2] and ∴ ↑ [H^+] found there

$$acids = H^+ \text{ donors} \quad bases = H^+ \text{ acceptors}$$

buffers - conjugate acid/base pairs that resist/minimize Δ pH

1- H_2CO_3/HCO_3^- - important plasma buffer

H_2CO_3 acts as an acid → neutralizes excess base

HCO_3^- acts as a base → neutralizes excess acid

2- protein buffer system - accounts for ~ 75% of buffering capacity

aa contain NH_3^+ group (H^+ acceptor) & COO^- (H^+ donor) at physiological pH

3- $H_2PO_4^-/HPO_4^{2-}$ - important intracellular buffer, primarily used by kidney

acid-base regulation involves:

    respiratory mechanism - provides rapid response
    urinary mechanism - slower acting but greater capacity to regulate pH

respiratory:

↑ pH (↓ H^+) → ↓ respiration rate/depth → ↑ CO_2 accumulation → ↑ H^+

↓ pH (↑ H^+) → ↑ respiration rate/depth → excess CO_2 elimination → ↓ CO_2

renal:

↑ pH (↓ H^+) → ↓ H^+ secretion, ↓ HCO_3^- reabsorption → ↓ pH as H^+ increases

↓ pH (↑ H^+) → ↑ H^+ secretion, ↑ HCO_3^- reabsorption → ↑ pH as H^+ decreases

	Blood	Tubule Cell	Urine	
reabsorbed, buffers → excess H^+	(HCO₃⁻) ← ⎯ CO_2	← HCO_3^- H^+ → H_2CO_3 ↑ $CO_2 + H_2O$	(H⁺) ← ⎯	secreted

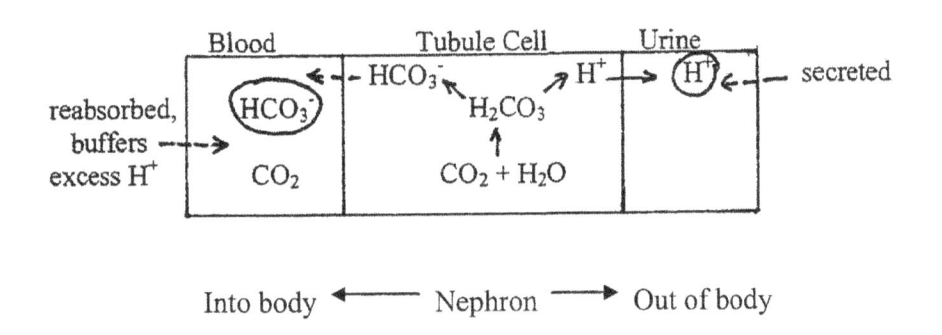

Into body ← Nephron → Out of body

acidosis, pH below 7.35 <

        respiratory - due to ↓CO_2 elimination (hypoventilation)

        metabolic - due to ↑ HCO_3^- loss or inadequate O_2 supply
            (vomiting lower GI, diarrhea) (↑ lactic acid)

alkalosis, pH above 7.45 <

        respiratory - due to hyperventilation, ↑ CO_2 elimination

        metabolic - due to excess H^+ loss/reabsorbing ↑ $[HCO_3^-]$
            (vomiting stomach contents, ↓ urine pH)

Chapter 28 - The Reproductive System

Male reproduction - produces sperm; transfers sperm to female; produces testosterone

spermatozoa - temperature sensitive, develop outside body within scrotum

cryptorchidism - failure of one or both testicles to descend into scrotum

testes - male gonads contain:
    seminiferous tubules - site of sperm production
    interstitial cells of Leydig - produce testosterone

– develop within abdominal cavity and move through **inguinal canal** during 7-8 mo of fetal development

Spermatogenesis - spermatozoa production within seminiferous tubules (ST);
– begins at puberty; several hundred million/day produced

ST contain 2 types of cells:
1) **Sertoli cells ("nurse" cells)** - nourish germ cells, produce some hormones
    – form **blood-testes barrier** - protect sperm against immune system attack
    (sperm are 1N, 23 chromosomes, instead of 2N, 46 chromosomes)

2) **Germ cells**

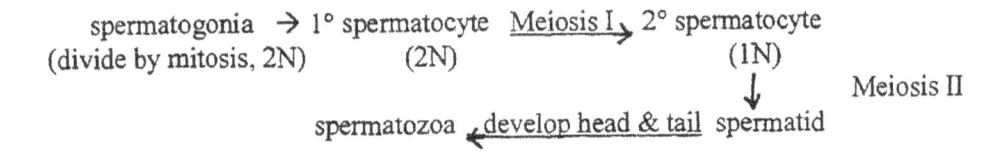

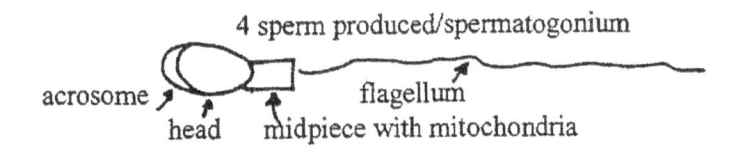

spermatogonia → 1° spermatocyte <u>Meiosis I</u> → 2° spermatocyte
(divide by mitosis, 2N)   (2N)   (1N)
                                             ↓ Meiosis II
         spermatozoa ← <u>develop head & tail</u> spermatid

4 sperm produced/spermatogonium

acrosome, head, midpiece with mitochondria, flagellum

drainage sequence for produced sperm:

seminiferous tubules → rete testis → vas efferentia → epididymus

epididymus - site of maturation & sperm storage before ejaculation

vas deferens - conveyed with spermatic cord through inguinal canal into pelvic cavity
– transports sperm from epididymus → urethra
– peristalsis helps move sperm through the duct
– severed in vasectomy

ejaculatory duct - formed from vas deferens + duct draining seminal vesicles

urethra - prostatic → membranous → spongy sections
 – passageway for urine & semen

semen = sperm + reproductive gland secretions (2-3 ml)
 – testes contribute spermatozoa to semen, represents ~5% of total semen volume

Seminal glands include:
1- **seminal vesicles** - ~60% of semen volume
 – fluid contains fructose to nourish sperm, PGs & fibrinogen

2- **prostate gland** - ~30% of semen volume
 – alkaline fluid neutralizes acidic vagina, also includes clotting factors

3- **bulbourethral (Cowper's) gland** - ~5% of semen volume
 – alkaline pre-ejaculate

coagulation function of semen - semen coagulates temporarily within the vagina after ejaculation
 – later, sperm become mobile again & swim onward
arousal - requires parasympathetic division ANS
emission - discharge of semen from seminal glands, requires sympathetic division ANS
ejaculation - expulsion of semen from urethra
resolution - after ejaculation, penis flaccid

normal sperm count in semen ≈ 75 - 400 million/ml semen
 (usually ~ 2-5 ml semen/ejaculate)

impotence - inability to achieve/maintain erection (physical vs. psychological causes)

during erection - arteries within erectile tissues dilate → engorgement with blood
 – veins are compressed; blood cannot drain
 – results in vasocongestion

male hormones -
+ at puberty, hypothalamus begins producing GnRH <u>anterior pituitary</u> FSH & LH
+ **FSH** - binds to Sertoli cells in seminiferous tubules, promotes spermatogenesis
+ **LH** - binds to Leydig cells, ↑ testosterone synthesis
+ **testosterone** - 1° male sex hormone
 – necessary for spermatogenesis, maintenance of 2° sexual traits
 – ↑ protein synthesis (generally ↑ metabolic rate)
 – stimulates hair growth

+ **inhibin** - from testes, inhibits FSH secretion (- feedback)

Female reproduction - produces ovum, female gamete
- produces female sex hormones
- houses developing fetus
- nourishes infant

ovary (with follicles) → fallopian (uterine) tube → uterus → vagina

oogonium <u>begins Meiosis I</u> → 1° oocyte (2 million @ birth; 300,000 left @ puberty)
 stops @ Prophase I

Meiosis I continues

ovum ← 2° oocyte completes development only <u>Meiosis II</u> 2° oocyte
 if fertilized stops @ Metaphase I

primary follicle - primary oocyte surrounded by granulosa cells

every ~ 28 days: 1° follicle → 2° follicle → Graafian follicle (fluid filled with theca interna/externa)

ovulation - release of 2° oocyte from Graafian follicle

corpus luteum - follicle development into a secretory unit after ovulation
- produces progesterone & estrogen
- degenerates within 10-12 days without pregnancy & becomes **corpus albicans**

1 ovum produced/oogonium + 3 polar bodies

with puberty, begin cyclic release of GnRH → LH, FSH
at level of ovaries & uterus, this results in monthly menstral cycle:
 1- follicle development & ovulation
 2- monthly changes in estrogen/progesterone secretion
 3- uterine lining (endometrium) changes

menstral cycle ≅ 28 days; **mensus** = sloughing/expulsion of endometrium

      day 1 - 5 mensus
         6 - 14 follicular/proliferative phase - "ovarian cycle"
         14 ovulation
         15 - 28 luteal/secretory phase - "uterine cycle"

during follicle development, GnRH → LH, FSH

LH/FSH - affect development of follicle within ovary
– both hormones peak in their concentration with a "surge" in production at ~ day 14 = ovulation signal

during "surge" - [LH] > [FSH];   **FSH** - 1° effect on granulosa cells
LH - initial effect on theca cells, later effects granulosa cells

+ also see rise in [estrogen] with peak prior to ovulation (this estrogen is produced by theca cells of follicle)
+ uterine mucosa begins to proliferate with increased [estrogen]
+ small rise in progesterone levels also seen

LH surge initiates ovulation & causes follicle to become corpus luteum after ovulation
LH causes 1° oocyte to complete Meiosis I
LH causes inflammatory response within follicle that leads to ovulation

with corpus luteum development, [estrogen, progesterone] >> preovulation levels
– also see decrease in GnRH, LH & FSH production after ovulation

with fertilization, **HCG** (human chorionic gonadotropin) is produced by the embryo & maintains the corpus luteum (CL) until placenta is functional

without fertilization, HCG is not produced & the CL degrades after day 25-26
– with ↓ [progesterone, estrogen], uterine lining degrades

uterine cycle - changes in uterine lining during follicular & luteal phases
by day 21 - lining ready for embryo implantation

estrogen - from corpus luteum, causes endometrial cells to proliferate
– stimulates production of progesterone receptors within uterine walls

progesterone - from CL, causes hypertrophy of endometrium

↓ [progesterone] → ↑ inflammatory substances → uterine contractions & expulsion of lining

PMS -(premenstral syndrome) - days 25-28, associated with rapid decline of female steroid hormones

Fertilization - occurs in upper 1/3 (ampulla) of fallopian tube
– oocyte viable for ~ 24 hrs after ovulation
– sperm viable for 48 - 72 hrs after ovulation within female reproductive tract
∴ 3 days before → 1 day after ovulation constitutes fertilization window (day 11-15)

lactation - milk production by mammary glands requires 3 hormones:
  estrogen - 1° responsibility for breast development during pregnancy
  prolactin - responsible for milk production
  oxytocin - responsible for milk letdown
– ↓ GnRH release by hypothalamus ("natural" birth control)

menopause - cessation of menstruation between ~ 40-50 yrs; "climacteric" - last mensus
"hot flashes" - due to ↓ estrogen levels

fertilization - sperm enters 2° oocyte → **zygote**

zygote - fertilized ovum
blastocyst - within 3-4 days after fertilization, zygote becomes fluid-filled cell mass
– stage that implants within uterine lining
  blastocoel - fluid-filled cavity
  inner cell mass - tissue from which embryo will develop
  trophoblast - surrounds blastocoel
      – secretes enzymes that digest endometrial cells of uterus & allows blastocyst to implant
      – secretes HCG
      – placenta forms from trophoblast

Prenatal period - conception → birth

  germinal period - first 2 weeks, germ layers formed
  embryonic period - 2nd - 8th week, most organ systems develop
  fetal period - last 7 months, "growing phase"

Control of pregnancy -
  sterilization - vasectomy & tubal ligation
  behavioral - abstinence, coitus interruptus, rhythm method
  barrier methods - mechanical - condom, diaphragm
                    chemical - spermicidal agents
  chemical methods - oral contraceptives, ↓ LH/FSH ∴ no ovulation
    ↑ estrogen/progesterone - uterine development (↓ GnRH, LH/FSH)
  IUD - prevents implantation

Appendix - Review Sheets

Review Sheet - Chapters 1 - 3

Chapter 1

Anatomy - gross, histology, cytology;   Physiology

Characteristics of living things - movement, metabolism (catabolism/anabolism), responsiveness, growth/development, reproduction, adaptation, organization

Organizational levels: chemical → organelle → cell → tissue → organ → organ system → organism

Overview of organ systems - know overall function of each organ system (Fig. 1.3)

Homeostasis - stimulus (stressor), receptor, control center, effector, response
  dynamic equilibrium, set point

negative feedback vs. positive feedback

body plan - bilateral symmetry, anatomical position; body planes - sagittal, transverse, frontal

anatomical comparatives: superior/inferior; anterior (ventral)/posterior (dorsal); medial/lateral; proximal/distal; superficial deep; parietal/visceral

body cavities - dorsal - cranial, vertebral cavities; ventral - thoracic (pleural, pericardial), mediastimum; abdominopelvic (abdominal, pelvic) - peritoneal

body regions - (Fig. 1.10, 1.11)

Chapter 2

Atomic structure: nucleus [proton (+, 1 amu) + neutron (0, 1 amu)], electron (-, 0 amu)

Atomic no. = #p = #e;   Atomic mass = #p + #n

isotopes - differ in #n; same #p,e;   importance of radioisotopes

Chemical bonds - (T 2-4); valence electrons determine chemical properties

ionic bonds: electron donor - metal, form cations (+); electron acceptor - nonmetal, form anions (-)
  formed between oppositely charged cations/anions

covalent bonds: sharing of electron pairs; nonpolar - equal sharing (H_2, O_2, N_2);
  polar - unequal sharing (H_2O)

hydrogen bonds: weak interaction between H and O/N; importance in biology

water - intramolecular bonds = polar covalent bonds; intermolecular bonds = H bonds
 important solvent (participates in solutions, suspensions, colloids); high specific heat; lubrication;
 participates in condensation/hydrolysis rxns

Chemical reactions :   reactants → products
  synthesis rxns - anabolic, endergonic; decomposition rxns - catabolic, exergonic; exchange rxns -
  neutralization rxn; redox rxns - oxidation = loss of e^-, reduction = gain of e^-

Reaction rate - incs. with [R], T, enzymes

Enzymes = biological catalysts, decrease activation energy for a reaction

Energy - ability to do work;   PE - stored energy, KE - energy of motion, heat - 'lost' energy

Acids - proton donors, produce H^+;   Bases - proton acceptors, produce OH^-
pH scale -   pH 7 = neutral $[H^+] = [OH^-]$
                        pH less than 7 = acidic $[H^+] > [OH^-]$
                        pH greater than 7 = basic $[H^+] < [OH^-]$

buffers - resist pH changes, eg. H_2CO_3/HCO_3^- ; neutralize excess base/neutralize excess acid

Carbohydrates - contain C, H, O in 1:2:1 ratio
   monosaccharides - simple sugars, polar
     pentoses - 5C sugars, found in NA; deoxyribose (DNA), ribose (RNA, ATP)
     hexoses - 6C sugars, cellular fuel sources; glucose, galactose, fructose

   disaccharides - 2 simple sugars joined w/ glycosidic (polar covalent) bond
     sucrose = glucose + fructose; maltose = 2 glucose; lactose = glucose + galactose

   polysaccharides - many repeating sugar units, no precise mw
     glycogen - animal storage; starch - plant storage; cellulose - plant CW, structural

Lipids - contain C,H,O (P)
   neutral fats - fuel source; glycerol + 3 FA (= triglycerides)
     saturated: C-C, solids, animal products, "unhealthy"
     unsaturated: C=C, liquid oils, plant products

   phospholipids - contain phosphate/N-base; found in CM
     hydrocarbon (FA) tail, hydrophobic, nonpolar; polar head, hydrophilic

   steroids - 4 interlocking rings, e.g. - vit D, bile salts, cholesterol, steroid hormones

Nucleic acids - information storage; made up of nucleotides (sugar, phosphate, N-base)
   DNA = double-stranded helix held together with H bonds, S(deoxyribose)-P covalent backbone
     A & G = purines;   C,T = pyrimidines;   A:::T and G:::C
   RNA = single stranded helix; A, G, C, U bases; ribose sugar;   mRNA, rRNA, tRNA
   ATP = A, ribose & 3 Phosphate groups; "energy currency", energy stored in P ~ P bonds

Proteins - aa bound by peptide bonds (polar covalent); contain C,H,O,N,S
   essential vs. nonessential aa
   proteins used for structure, transport, enzymes, hormones, Ab
   1°, 2°,3°,4° structure; imp. of H bonds; α helix, β pleated; fibrous vs. globular proteins

   enzymes: -ase ending, specificity, lock & key model of action, active site, cofactors/coenzymes

Chapter 3

cell membrane - (F 3-2), fluid mosaic model, phospholipid bilayer + cholesterol + proteins
integral vs. peripheral proteins; glycoproteins (marker, recognition fcn)

cell organelles - structure and function (T 3-1)

movement across CM - passive vs. active processes   (T 3-2)

diffusion, osmosis (isotonic, hypotonic, hypertonic), filtration, facilitated diffusion
active transport, cotransport, endocytosis (phagocytosis/pinocytosis), exocytosis

cell metabolism - aerobic/anaerobic respiration, protein synthesis (DNA → RNA → protein) - transcription, translation, DNA replication

cell cycle – interphase (G_1, S, G_2 phases), mitosis (prophase, metaphase, anaphase, telophase), meiosis (I&II); mitotic spindle, crossing over, cytokinesis, chromatid, centromere (Figs.3.34, 3.35); comparison of mitosis & meiosis (T3.3)

Review - Chapters 4-8

Chapter 4-

Epithelial tissue - functions; avascular; no matrix; basement membrane; free border; tight-fitting junctions
  shapes - squamous, cuboidal, columnar
  organizational complexity - simple, stratified, pseudostratified, transitional

  Know names, locations & functions:
  simple squamous/cuboidal/columnar - diffusion, secretion, absorption
  stratified squamous - moist or keratinized; protection
  transitional - stretch without tearing

  specialized features - goblet cells, cilia, microvilli, keratin

  cell connections - desmosomes/hemidesmosomes; tight jcns (zona adherens/occludens); gap jcns

  glandular epithelium - exocrine vs. endocrine; unicellular (Goblet) vs. multicellular; simple vs. compound; tubular/acinar/alveolar; mucous /serous; merocrine/apocrine/holocrine glands

Connective tissue - nonliving matrix with protein fibers (collagen, elastic, reticular)
  -blasts, -cytes, -clasts; ground substance (hyaluronic acid, proteoglycan)

  Know names, locations & functions:
    1- connective tissue proper -
    loose (areolar) connective tissue - pkging/space-filling fcn
    dense connective tissue - regular (tendons/ligaments) vs. irregular (dermis)
    2- adipose tissue - adipocyte; fat storage, insulation, protection fcns.
3- reticular connective tissue
4- bone marrow - yellow (fat storage); red (blood cell production)
    5- cartilage - chondrocytes, lacunae, perichondrium, avascular, semirigid matrix
        hyaline, fibro-, elastic cartilages (know distribution/fcn differences)
    6- bone - Haversian system (osteon), Haversion canal, canaliculi, osteocytes, lacunae, lamellae, periosteum
        organic matrix = collagen, ground substance → flexible strength
        inorganic matrix = Ca/P salts (hydroxyapatite crystals) → rigidity, hardness
    7- blood - erythrocytes, leukocytes, thrombocytes

Muscle tissue - skeletal/cardiac/smooth muscles
  Know location, control, fiber shape, # nuclei, striations

Nervous tissue - neurons, glial cells, cell body, axon, dendrites

Embryologic germ layers - ecto-/meso-/endoderm

Membranes - mucous/serous/synovial; parietal vs. visceral; peritoneum/pleura/pericardium

Tissue repair - regeneration/replacement; labile/stable/permanent cells
Wound repair - clot → scab → granulation tissue → permanent tissue (scar)
Inflammatory response - redness, swelling, heat, pain, loss of fcn

Chapter 5-

Functions of integument - protection, T regulation (evaporative vs. radiative cooling), vit D, sensation, excretion
Epidermis vs. dermis vs. hypodermis (subcutaneous layer, superficial fascia)
Keratinocytes, melanocytes, Langerhans cells

Epidermis- stratum basale - mitotic, keratinocytes/melanocytes found here
stratum spinosum - prickle-cell layer, desmosomes
stratum granulosum - keratinohyalin present
stratum lucidum - soles & palms
stratum corneum - keratin, desquamation

thin skin vs. thick skin; keratinocytes → keratin (water-proofing protein); melanocytes → melanin within melanosomes (absorb UV radiation)
skin color - melanin, carotene, blood supply; albinism, cyanosis

Dermis - papillary layer - dermal papillae = fingerprints, grasping; superficial blood supply
reticular layer - irrgeular connective tissue

cleavage/tension lines; stretch marks
burns - first/second/third degree; % integument involvement

Hair - shaft, root, hair follicle, hair papillae, hair bulb; arrector pili (smooth muscle); growth vs. resting stages; melanin → color

Nails - stratum corneum; protection/manipulation fcn

Sebaceous (oil) glands - sebum, lubricates hair, holocrine gland associated with hair follicle

Sudoriferous (sweat) glands - eccrine (diffuse, small pore) vs. apocrine (axilla/groin, large pore)

Chapter 6 -

Skeletal system functions - support, protection, movement, storage (Ca/P, fat), blood cell production

Bone shapes - short, long, flat, irregular
Compact (dense) vs. Spongy (cancellous) bone

T6-1, F6-3 - diaphysis, epiphysis, epiphyseal plate, trabeculae, periosteum (fibrous/osteogenic), marrow (medullary) cavity, yellow/red marrow, endosteum, articular cartilage

F6-10 - compact bone structure - Haversion system; canaliculi, Haversian canal, Volkman's canal

osteoblasts → osteocytes; osteoclasts → resorption
matrix - organic vs. inorganic materials

bone development - intramembranous (flat) vs. endochondral (long) ossification   T6-2
mesenchyme; woven vs. lamellar bone; appositional vs. interstitial growth

epiphyseal plate - zones of resting, proliferation, maturation, calcification   Fig. 6.15
bone growth - appositional vs. endochondral

fractures - complete/incomplete, simple/compound, comminuted, reduced, transverse/oblique/
    spiral, compaction, compression
  clot (procallus) → fibrocartilage callus → osseous callus

tendons = muscle → bone attachment; high collagen content, flexible strength
ligaments = bone → bone attachment; reduce range of motion around joint

hormonal/nutritional influences on bone growth - vit D/C; GH; steroids; thyroxine;
    PTH; calcitonin
Calcium homeostasis - osteoblasts/-clasts; PTH/calcitonin

diseases - scurvey, rickets, osteomalacia, osteomyelitis, osteoporosis, pituitary giant/dwarf,
    arthritis - osteo-, rheumatoid

Chapter 7 -

axial vs. appendicular skeleton
bones of: orbit of eye; nasal cavity; nasal septum; hard palate; with paranasal sinuses
spinal curves - primary vs. secondary
intervertebral discs - annulus fibrosus/nucleus pulposus; herniated discs
lordosis, kyphosis, scoliosis
differences between male/female pelvis; pelvic inlet/outlet   Fig. 7.38, T 7-8

Chapter 8 -

Types of joints = fibrous, cartilaginous, synovial

Fibrous - sutures (fontanels), syndesmoses, gomphoses

Cartilaginous - synchondroses, symphysis

Synovial - joint capsule (fibrous capsule & synovial membrane); synovial fluid - lubricating, nutritive;
    articular cartilage; articular discs (menisci); bursa

Movement types - mono-, bi-, multiaxial

Structural types - know examples   Fig. 8.8-8.13, T 8-2
    plane/gliding; saddle; pivot; hinge; ball-and-socket; ellipsoidal

Types of movement – Fig. 8.14-8.26
    flexion/extension; abduction/adduction; rotation; pronation/supination; circumduction;
    elevation/depression; protraction/retraction; lateral/medial excursion; opposition/reposition;
    inversion/eversion; dorsiflexion/plantar flexion

Special features of joints -
TMJ - functional dislocation
shoulder vs. hip - range of motion vs. ease of dislocation
knee - menisci, collateral ligaments, cruciate ligaments; flexion/extension with rotation
ankle - talocrural jt. - modified hinge
arches of foot - distribute body weight

Review Sheet - Ch. 9-11

Chapter 9

muscle - contractility, excitability, extensibility, elasticity; compare skeletal/smooth/cardiac (T 9.1)
muscle cell = muscle fiber; myoblast
sarcolemma, external lamina, endomysium, perimysium, epimysium, fascia

myofilament - actin (thin) = actin + troponin + tropomyosin   (importance/role of each)
       myosin = rod + flexible head (w/ ATP)
sarcomere - I/A bands, Z-line, H zone, M line
T-tubules, sarcoplasmic reticulum
Sliding filament theory - cross-bridge formation/release

Neuromuscular jcn (Fig. 9.11-9.12) - synaptic vesicles, pre/postsynaptic sides, synaptic cleft, receptors
   neurotransmitter (Ach); role of Ca/Na ion movement; Achase

Excitation-contraction coupling - AP → sliding filament events, Fig. 9-6,9-14,9-15
power vs. recovery stroke
muscle relaxation events (Ca reabsorption)

muscle twitch - lag, contraction, relaxation phases; electrochemical vs. mechanical events
  (Fig. 9-16, T 9-2)   individual fibers
subthreshold, threshold, above threshold stimulus
motor unit - size related to activity/strength
graded response - whole muscle phenomenon
maximal stimulus, recruitment
incomplete/complete tetanus; multiple wave summation;   treppe

isometric vs. isotonic contractions;   muscle tone;   (T 9-3)
concentric vs. eccentric contractions
active tension, passive tension
fatigue - pscyhological, muscular, synaptic; rigor mortis; physiological contracture

anaerobic vs. aerobic respiration; oxygen debt; lactic acid accumulation; creatine phosphate
slow-twitch (endurance) vs. fast-twitch (quick burst of speed) fibers
exercise → hypertrophy; aerobic vs. intensity training;   disuse → atrophy
shivering = heat production, homeostasis

smooth muscle - non-striated, importance of Ca/calmodulin; myosin kinase/phosphatase
  visceral vs. multiunit smooth muscle; RMP ~ -55-60 meV; epinephrine/oxytocin influence
  constant tension, innervated by ANS, contracts when stretched

cardiac muscle - autorhythmic, intercalated discs

spastic vs. flaccid paralysis; myasthenia gravis; anabolic steroids; poliomyelitis;
muscular dystrophy (Duchenne/Facioscapulohumeral); fibrosis, fibrositis, cramps, fibromyalgia

Chapter 10

tendons - cordlike vs. aponeurosis
muscles - origin, insertion, belly; prime mover, synergist, antagonist, fixator (e.g.)
classification of muscles
contraction - force, lever, fulcrum, resistance; Class I/II/III levers

Chapter 11

CNS (brain, spinal cord) vs. PNS (nerves [axons], ganglia [cell bodies])
PNS divisions - afferent (sensory), efferent (motor)
efferent divisions - somatic (SNS), voluntary; autonomic (ANS), involuntary
ANS divisions - sympathetic (stress mobilization), parasympathetic (resting/return to normal)

neuron structure/fcn - (Fig.11.4) - cell body, dendrites, axon, Nissl substances, axon hillock,
   axolemma, telodendria, terminal boutons, myelin sheath, nodes of Ranvier

neuron type: afferent, efferent, association (interneuron)
          multipolar, bipolar, unipolar

neuroglia: astrocytes (blood-brain barrier); ependymal cells (ventricles, choroid plexus);
  microglia (defense, phagocytic); Schwann cells (produce myelin w/in PNS);
  oligodendrites (produce myelin w/in CNS); satellite cells (surround ganglia)

unmyelinate/myelinated axons; Type A/B/C fibers; saltation

white matter vs. gray matter; nerve vs. nerve tract; ganglia vs. cortex/nuclei

neurilemma, endoneurium, perineurium (fascicle), epineurium (Fig. 12.12)

RMP +++++++   location of Na/K/Cl/proteins at rest; Na/K pump importance
      -----------   - 85 meV;
      +++++++   effect of membrane permeability changes on RMP
(T 11-2)
polarized/hypopolarized (depolarized)/ hyperpolarized
ion channels - voltage-sensitive/ligand-gated
   -Na channels open 1st, close 1st; K channels open more slowly, stay open longer
   - direction of Na/K movement when channels open vs. direction of active transport
local potential, graded potential, summation, propagation T11-3, Fig. 11.24
action potential - 'all-or-none'; depolarization/repolarization phases, after potential (Fig. 11.20; T11-4)

absolute/relative refractory period Fig. 11.22
synapse - electrical vs. chemical, neurotransmitter (Fig. 11.27, 11.28)
accomodation

nerve-nerve synapse - compare synapse action of Ach/NE (MAO)
neurotransmitters/neuromodulators (T11-5)

EPSPs, IPSPs, axo-axionic synapses (presynaptic inhibition/facilitation)
summation - temporal vs. spatial (Fig. 11.31)

reflexes - receptor → sensory neuron → association neuron → motor neuron → effector

pathways - convergent vs. divergent, oscillating circuits (afterdischarge)

Review Sheet - Chapters 12-16

Chapter 12

Spinal cord - gray matter/inside (horns); white matter/outside (ascending/descending tracts)
   31 pr spinal nerves; conus medullaris, cauda equina, filum terminale;
   x-section (Fig. 12.3) - dorsal root w/ ganglion = sensory; ventral root (motor); central canal w/ CSF
spinal reflexes - stretch (patellar) [no association neuron]; withdrawl [association neuron present]

Spinal nerves - C1-C8; T1-T12; L1-L5; S1-S5; Co1; dermatomes; dorsal/ventral/sympathetic rami
Plexuses: cervical (C1-C4, phrenic n.); brachial (C5-T1, axillary, radial, ulnar n.;
   lumbar (L1-L4, obturator, femoral n.); sacral (L4-S4, sciatic = ischiadic → tibial + fibular n);
   coccygeal (S4,5-Co1); intercostal n. (T2-T12, don't form plexus)

Chapter 13

gray matter = nucleus, cortex; white matter = nerve tract
brainstem = pons, medulla, midbrain; reticular formation (RAS)
medulla - vital centers, pyramids, decussation, olives; pons - relay (cerebellum/cerebrum), reflex centers
midbrain - tectum, corpora quadrigemina; superior colliculi = visual reflexes, inferior colliculi = hearing
   tegmentum w/ red muclei; cerebral peduncles; substantia nigra
diencephalon = thalamus, hypothalamus, sub/epithalamus
   thalamus - sensory relay center (except olfaction); intermediate mass, 3rd ventricle;
     lateral geniculate nucleus (vision); mediate geniculate nucleus (auditory); ventral posterior nucleus
     (other sensory); anterior/medial nuclei (limbic system, prefrontal cortex); ventral anterior/lateral
     nuclei (basal ganglia, motor cortex)
   hypothalamus - controls pituitary gland (attached via infundibulum); mamillary bodies (olfaction);
     homeostatic fcns, body T regulation (T13-2)
   pineal gland - melatonin
cerebellum - controls muscle movements/tone; "comparator" fcn; folia, arbor vitae;
   flocculonodular lobe (balance/tone); vermis (gross motor control); lateral hemispheres (fine motor
   control- smooth, flowing movt); field-sobriety tests
cerebrum - gyrus, sulcus, fissure, R/L hemispheres, lobes: frontal (motor, motivation, mood, aggression);
   parietal (sensory); occipital (vision); temporal (auditory, abstract thought/judgement); insula (deep);

PNS - cranial = 12 pr; spinal = 31 pr; fcn = sensory, somatic motor, parasympathetic

Know cranial nerves by name/#; distribution/function; sensory/motor/both/parasympathetic
 (T13-4, 13-5)

Chapter 14

general vs. special senses; mechano-, chemo-, photo-, thermo-receptors, nociceptors;
5 steps in sensation; extero-, viscero-receptors, proprioceptors
afferent nerve endings - T 15-2, F 15-1, free nerve endings, cold/warm/pain receptors, Merkel's disc,
   hair follicle receptors, Pacinian corpuscles, Meissner's corpuscles, Ruffini's end organ,
   Golgi tendon organ, muscle spindle fibers

  limbic (part of limbic syst); R/L cerebral dominance - motor (opposite); spatial/analytical fcns.
   fibers - association/commissure/projection
   postcentral gyrus - 1° somesthetic/sensory area; P,T,pain info.; sensory homunculus
   precentral gyrus - 1° motor cortex; motor homunculus; premotor (planning/staging);
     prefrontal (motivation; emotional behavior)

speech - Wernicke's vs. Broca's areas
brain waves - EEG: $\alpha, \beta, \delta, \theta$ components
basal ganglia = subthalamic nuclei, substantia nigra, corpus striatum (caudate/lentiform nuclei);
   dec. muscle tone, eliminates unwanted muscle activity
limbic system = cerebral cortex (hippocampus), basal ganglia, thalamus/hypothalamus, olfactory bulbs,
   fornix; influences emotions, moods, pain/pleasure, survival skills
paleocortex vs. neocortex

spinal tracts – T14-3, 14-4; ascending - spinothalamic, spinocerebellar;
   descending - pyramidal (conscious movts. - corticospinal, corticobulbar);
         extrapyramidal (unconscious movts. - vestibulospinal, tectospinal, reticulospinal)
proprioception
pain - somatic, visceral, referred, phantom, chronic
meninges - dura mater, arachnoid layer, pia mater; epidural, subdural subarachnoid spaces
CSF - extracellular fluid, formed by choroid plexus of ventricles; arachnoid granulations
ventricles - lateral (septum pellucidum), 3rd ventricle, cerebral aqueduct, 4th ventricle, central canal
   (spinal cord)
brain protectors = cranium, dura mater, CSF (shock absorber, cushioning medium)
CNS pathologies - apraxia, aphasia, dyskinesia

Chapter 15

olfaction - bipolar neurons, chemoreception, quick adaptation, high sensitivity,
   olfactory vesicle/cilia; no thalamic synapse

gustation - taste buds/cells/hairs/pores; papillae - fungiform, filiform, foliate, circumvallate
   salty, sweet, sour & bitter discrimination; involvement of olfaction in taste
   CN VII,IX,X → medulla → thalamus → gustatory cortex (postcentral gyrus)

vision - layers of eye & their fcns
   fibrous tunic - cornea (refraction, light entrance); sclera (protection, muscle attachment)
   vascular tunic - choroid (melanin/absorb stray light, nutrient supply); ciliary muscle (lens shape);
      iris (controls pupil size: circular/constriction/parasympathetic; radial/dilate/sympathetic)
   retina - pigmented layer; nervous layer (rods/cones → bipolar neurons → ganglionic neurons)

   anterior compartment (aqueous humor); posterior compartment (vitreous humor)
   lens - biconvex, accomodation (flat = distance; plump = near), focal point (converge/cross)
      inverted image @ retina

refraction - cornea → aqueous humor → lens → vitreous humor
reflection - stray light
convergence - medial rectus
emmetropia – myopia – hyperopia – presbyopia
visual fied overlap & binocular vision
diseases of eye

rods/cones → bipolar neurons → ganglion neurons → CN II → optic chiasma → optic tract →
lateral geniculate nucleus (thalamus) → visual cortex (occipital lobe)

rods - periphery, shape, dim light (night vision); rhodopsin = retinal + opsin
   light causes bleaching of rhodopsin & hyperpolarization of rod CM; light/dark adaptation
cones - center (fovea centralis/macula lutea), acuity, color vision, red/green/blue cones
   iodopsin, sex-linked characteristic

auditory system - outer ear (auricle, extl auditory canal, tympanic membrane);
  middle ear (malleus, incus, stapes; auditory canal, oval window); inner ear (cochlea)

cochlea - scala vestibuli, scala tympani (w/ perilymph), scala media/cochlear duct (w/ endolymph);
  organ of Corti = hair cells + tectorial membrane   (T 15-2, Fig. 15-25)
loudness = wave amplitude, dB; pitch = frequency, cps/Hz

CN VIII → medulla → inferior colliculi → medial geniculate nucleus (thalamus) →
 auditory cortex (temporal lobe)

balance - static equil. (vestibule = utricle + saccule); dynamic equil. (semicircular canals)
vestibule - macula w/ hair cells & otoliths; bending of hairs causes depolarization
semicircular canals - ampula w/ crista (hair cells w/ cupula extending over them)
  movement of endolymph causes depolarization

ear disorders - tinnitus, motion/space sickness, otitis media

Chapter 16

compare somatic vs. autonomic NS   T 16-1
pre/postganglionic neurons; autonomic ganglia
compare sympathetic (thoracolumbar) vs. parasympathetic (craniosacral) divisions T 16-2
sympathetic chain ganglia, white/gray ramus communicans, spinal/sympathetic/splanchnic nerves
adrenal medulla (epi/norepi)

adrenergic vs. cholinergic neurons   (NT release)
cholinergic receptors (nicotinic, muscarinic); adrenergic receptors (α, β)

stimulatory/inhibitory effects, dual innervation, opposite effects, cooperative effects,
general/localized effects, fcns at rest/activity

Review Sheet - Chapters 17,18,19

Chapter 17

nervous/endocrine comparisons
hormones (chemical messengers), neurohormones, neurotransmitters, auto/paracrine, neuromodulators
amplitude- vs. frequency-modulated signals
chemical structure - aa derivatives, peptides, proteins, lipid-soluble (steroids, PGs)
secretion regulation - [metabolite], nervous system, another hormone (tropic)
hormone transport - bound to plasma proteins
half-life of hormones - factors involved
membrane-bound receptors (protein hormones); hormone = 1st messenger; H-receptor → 2nd messenger
G-protein (α,β,γ); α-GTP= alter CM permeability; alter enzyme activity; inc./dec. [2nd messenger]
G-protein can open Ca^{+2} channels, inc. [cAMP/cGMP], inc. [DAG, IP_3]
cascade effect - amplification
down vs. up regulation
intracellular receptors (steroid, aa derivative hormones) - "turn-on" protein synthesis

Chapter 18

interaction between hypothalamus/anterior pituitary; hypothalamohypophyseal portal system/tract
pituitary - adenohypophysis vs. neurohypophysis
releasing/inhibiting hormones - names, targets, fcns (T18-1)
ADH -release conditions & effects (Osm/blood volume/BP) (Fig. 18.5); diabetes insipidus, EtOH/caffeine
oxytocin - targets & fcns
anterior pituitary hormones - (T 18-2)
GH - protein synthesis, fat breakdown, inc. blood glucose; release conditions & effects (Fig. 18.6)
   dwarfism, pygmy (- somatomedins), giantism, acromegaly
thyroid - follicles → T3/T4; parafollicular cells → calcitonin (dec. blood Ca^{+2})
   T3/T4 production - I pump, thyroglobulin - tyr; TBG carrier, T3 activity > T4 activity (Fig. 18.8)
   - inc. metabolic rate - "turn on" genes, inc. mitochondria activity (heat)
   - effects of hypo-/hypersecretion (T18-4); thyroid pathologies (T18-5)
parathyroid → PTH → inc. osteoclast activity (bone); inc. Ca^{+2} reabsorption (kidney);
   inc. vit D synthesis/ Ca^{+2} absorption (intestine)
   hypo-/hypercalcemia (T18-6)
adrenal medulla - epi/norepinephrine; fight-or-flight syndrome, inc. blood glucose (Fig. 18.13)
adrenal cortex- zona glomerulosa → mineralocorticoids (aldosterone);
   zona fasciculata → glucocorticoids (cortisol); zona reticularis → androgens
aldosterone - inc. Na^+/water reabsorption → inc. blood volume, BP; inc. K^+/H^+ excretion → alkalosis
cortisol - inc. blood glucose, inc. fat/protein brkdwn, dec. immune response (Fig. 18.14; T18-8,T18-9)
Cushing's syndrome vs. Addison's disease
pancreas - exocrine (acinar cells) vs. endocrine (Islets of Langerhans - α,β,δ cells)
glucagon - inc. blood glucose (glycogenolysis, gluconeogenesis); high conc. during fast   T 18-10
insulin - dec. blood glucose (inc. uptake into cells); high conc. after meal   (T18-11, Fig. 18.16, 18-17)
sympathetic NS = +glucagon/-insulin; parasympathetic NS= -glucagon/+insulin
somatostatin - dec. insulin/glucagon release
diabetes mellitus - Type I (juvenile, insulin-dependent); Type II (adult-onset, noninsulin-dependent)
insulin shock vs. diabetic coma   (clinical focus)
regulation of blood nutrients during exercise (Fig. 18.18)
stress response - stages 1,2,3

Chapter 19

- blood functions - transportation, maintenance, protection
- plasma - proteins (albumin, globulin, fibrinogens)
- erythrocytes, leukocytes, thrombocytes
- hematopoeisis (Fig. 19.2); erythropoeisis (Fig. 19.5) - erythropoeitin
- RBCs - biconcave, anucleate, #s, fcns (carbonic anhydrase activity)
- hemoglobin = heme + globin; oxy-, deoxy-, carboxy-, carbaminohemoglobin
- hemocytoblasts → proerythroblasts → basophilic erythroblasts → polychromatic erythroblasts → late erythroblasts → reticulocytes → erythrocytes
- blood disorders - clinical focus
- RBC recycling - biliverdin/bilirubin
- WBCs - nucleate, #s, agranulocytes vs. granulocytes; fcns of each type
- hemostasis - blood clotting, involves platelets + clotting factors/cofactors
- clot formation – Fig. 19.11 (know all imp. intermediates/factors); intrinsic vs. extrinsic pathways
- anticoagulants - heparin, antithrombin
- blood grouping - antigen/antibody reactions; agglutination; ABO blood group (donor/recipient)
- Rh blood group - erythroblastosis fetalis
- blood tests - CBC, hematocrit, cross-matching, [hemoglobin], WBC count (differential), platelet count, prothrombin time measurement

Review Sheet - Chapters 20,21,22

Chapter 20

heart - apex vs. base
fibrous pericardium, serous pericardium (parietal, visceral layers)
epicardium, myocardium, endocardium
R/L atria (with auricles) - receiving chambers; R/L ventricles - pumping chambers
atriventricular valves - tricuspid [R], bicuspid [L]; chordae tendinae, papillary muscles, cusps
semilunar valves - pulmonary [R], aortic [L]
arteries - carry blood away from heart; veins - carry blood towards heart
recognize branches of R/L coronary arteries (Fig. 20.6); coronary veins, coronary sinus
interatrial/interventricular septum
blood flow through the heart - know handout, oxygenated/deoxygenated condition of blood
autorhythmic conduction system - SA node (pacemaker), AV node (delay), bundle of His,
    R/L bundle branches, Purkinje fibers   (Fig. 20.13)
cardiac conduction potentials - **Fig. 20.14** - depolarization (Na^+ open); partial repolarization
    (K^+ open); plateau (K^+ closed, Ca^{+2} open); repolarization (Ca^{+2} closed, K^+ open); no tetany
    action of tetrodotoxin; Mn^{+2}/verapamil; epi/norepinephrine
ECG - P wave, QRS complex, T wave; electrical events/mechanical events that follow
cardiac cycle events - 0.8 sec/cycle ~ 72 beats/min
cardiac arrhythmias (T20-1)
systole vs. diastole *******Fig. 20.18, 20.19, T20-2******* interpret figure (P curves, Δ volume,
    valves open/close; stroke volume calculation; isometric contraction/relaxation;
    end diastolic volume; end systolic volume)
heart sounds - 1st, 2nd, 3rd; murmurs, stenosis
Cardiac output = heart rate x stroke volume ~ 5 L/min
cardiac reserve, maximum cardiac output
Blood Pressure = cardiac output x peripheral resistance
peripheral resistance increases w/ increasing viscosity & vessel length; decreases with increasing
    radius of vessel
intrinsic regulation of heart - venous return (Starling's law), preload, inc'd stretching of SA node
extrinsic regulation of heart - parasympathetic (vagus) - Ach, reduces HR (inc. K^+ permeability);
    sympathetic (cardiac nerves) - norepi, incs. HR & SV
baroreceptors - inc. BP → medulla → parasympathetic outflow → decrease HR, BP
  (stretch)   dec. BP → medulla → sympathetic outflow → increase HR, SV; increase BP
chemoreceptors - in medulla, sensitive to increased H^+/CO_2 → increased sympathetic outflow →
    increased HR & SV, also vasoconstriction
ion concentrations - effects of K^+/Ca^{+2}
effect of temperature

Chapter 21

systemic circulation (aorta → R atrium), pulmonary circulation (pulmonary trunk → L atrium),
    coronary circulation (L/R coronary a. → coronary sinus)
elastic arteries (aorta) → muscular arteries → arterioles → metarterioles → precapillary sphincter →
    capillary bed → venules → veins → vena cava
capillaries - endothelium, site of diffusion; fenestrated/sinusoidal/continuous; true vs. thoroughfare
    capillaries; slow blood flow, high x-sectl area
artery/vein wall - tunica intima (only layer in capillaries); tunica media (smooth muscle & elastic fibers -
    larger in arteries); tunica adventitia/externa (connective tissue - larger in veins)
veins - thin-walled, large lumen, valves, blood reservoirs, varicosity (toward heart)

arteries - thick-walled, high-pressure vessels (away from heart)

arterioles - largest ΔP, imp. in BP regulation
sympathetic innervation of blood vessels only → + = constriction; - = dilation; vascular tone
arteriosclerosis, atherosclerosis, angioplasty, coronary bypass surgery
flow patterns - laminar vs. turbulent (audible) blood flow
BP - measured w/ sphygmomanometer; Korotkoff sounds - 1st = systolic P; disappearance = diastolic P
   normal = 120 mmHg/80 mmHg; hypertension; pulse pressure
peripheral resistance - incs. with length of vessel & viscosity of blood; decs. with incd. vessel radius
critical closure P
Poiseuille's Law - flow varies directly w/ ΔP & r^4; varies inversely w/ v & l
Law of LaPlace - force = diameter X pressure; dec. P will ultimately collapse vessel
vascular compliance - greatest in veins (64% blood volume)
fluid exchange across capillary - hydrostatic P (filtration) - imp. @ arterial end, 30 mmHg;
  osmotic P (absorption) - imp. @ venous end, 10 mmHg   ***Fig. 21.32***
causes of edema (interstitial fluid accumulation)
control of blood flow to tissues - T21-15; local (O_2, CO_2, H^+), nervous system (sympathetic ANS),
  humoral (epi/norepi) control
mean arterial pressure (MAP) control
   short-term = baroreceptors, chemoreceptors, CNS ischemic response, adrenal medulla (epi)
   long-term = renin-angiotensin-aldosterone, vasopressin (ADH), ANF, fluid shift, stress relaxation
types/stages of shock; anastomosis; aneurysm

Chapter 22

lymp capillaries → lymph veins → lymph nodes → lymph veins → R lymphatic/thoracic ducts →
  R/L subclavian v.
functions - fluid balance, fat absorption (lacteals), defense
lymph nodules - tonsils (pharyngeal, palatine, lingual - filter air); lymph nodes - filter lymph;
spleen - red pulp (filters blood), white pulp (lymphocyte production)
thymus - T cell production; involution

immunity - nonspecific (innate) vs. specific (adaptive - possesses memory & specificity)
nonspecific processes - mechanical mechanisms, chemicals (complement, interferon, histamines,
   kinins, PGs, pyrogens, leukotrienes, surface chemicals - T22-1), cells (neutrophils, macrophages,
   basophils/mast cells/eosinophils, NK lymphocytes), inflammatory response (local vs. systemic)
specific immunity - cell-mediated (T cells) vs. antibody-mediated (B cells)   **T22-3**
   2nd responses are faster/stronger = memory
antigen (foreign vs. self), hapten; antibody - constant vs. variable regions, Ig G,M,A,E,D (T22-5)
lymphocyte activation - binds to antigenic determinant (epitope), MHC, costimulation (cytokines)
tolerance; role of regulatory T cells (helper T & suppressor T cells)
Ab complexes with Ag - interfere/inactivate; agglutination; opsonization; complement activation;
   inflammatory response
B cells display Ag determinants; helper T cells secrete IL-2; B cells develop into plasma & memory cells
   plasma cells → Ab, memory cells → facilitate 2nd exposure; primary vs. secondary responses
cell-mediated process - Ag activates effector T cells → memory T cells; cytotoxic T cells lyse cells,
   produce lymphokines; delayed hypersensitivity cells → allergic reactions
lymphokines - activate interferon, lymphotoxins, macrophage-activating factor, chemotaxis, IL-2
acquired immunity - active (natural vs. artificial); passive (natural vs. artificial); vaccination
monoclonal Abs, anaphylaxis, cytotoxic rxns, contact hypersensitivity, autoimmune diseases,
   immunodeficiency

Review Sheet - Chapters 23, 24 & 25

Chapter 23

Respiration = ventilation, external respiration, transport, internal respiration (see handout)
Nasal cavity → naso-, oro-, laryngo-pharynx → larynx → trachea → 1° bronchi → 2° bronchi →
3° bronchi → bronchioles → alveoli
functions of: nasal conchae, paranasal sinuses, mucus, cilia (beat direction in nasal cavity/trachea)
sneeze vs. cough
pleural cavity, parietal pleura, visceral pleura, pleural fluid (function)
larynx - fcn of epiglottis, vocal folds (expiration)
trachea - fcn of cartilage rings; mucociliary escalator
bronchi → bronchioles - reduce diameter, only smooth muscle left (asthma)
alveoli - site of external respiration, simple squamous epithelium, dust cells, fcn of surfactant (RDS)
muscles of respiration – Fig. 23.10; quiet vs. forced; active vs. passive
negative intrapleural pressure (pneumothorax)
respiratory volumes - TV, IRV, ERV, RV, IC, EC, FRC, VC, TLC, dead space, MRV, AVR (Fig. 23.15)
compliance, factors influencing air flow (thickness of respiratory membrane, surface area, ΔP, solubility)
oxyhemoglobin, reduced hemoglobin, carbaminohemoglobin, HCO_3^-, carbonic anhydrase, Cl^- shift
O_2-Hb dissociation/saturation curve – Fig. 23.17, 23.18; % satn @ 15, 40, 100 mmHg
nervous control of respiration - medulla (inspiratory/expiratory centers); pons (pneumotaxic, apneustic
centers), stretch receptors (Hering-Breuer reflex)
chemical control - central (medulla) : [CO_2/H^+], hypercapnia; peripheral (carotid/aorta) : hypoxia

Chapter 24

functions of digestive system - ingestion, propulsion, digestion (mechanical/chemical), absorption,
   excretion
tunics of GI tract - mucosa - epithelium (stratified squamous, → stomach; simple columnar, stomach →)
   lamina propria & muscularis mucosa; submucosa - CT; muscularis - skeletal, →1/3 esophagus;
   smooth, 2/3 esophagus →, circular, longitudinal (& oblique), myenteric plexus; serosa (adventitia) -
   visceral peritoneum.
functions of digestive organs - T24-1
oral cavity - matication, food → bolus; salivary glands - saliva (amylase, mucus, lysozyme, IgA);
deglutition - voluntary/involuntary phases; oro-, larygnopharynx, esophagus
esophagus - peristalsis, mucus, gravity aid in quick swallowing
stomach - bolus → chyme; mucus cells, parietal cells, chief cells; fcn of HCl, pepsin, intrinsic factor,
   mucus, rugae, gastrin, pyloric pump, mixing waves/peristalsis
small intestines - duodenum, jejunum, ileum; fcn of villi/microvilli & plicae circularis, segmentation,
   lacteals, secretin/CCK
liver - fcns include: bile secretion, glucose → glycogen, nutrient interconversion, detoxification,
   phagocytosis, storage, plasma protein synthesis
gallbladder - bile storage
pancreas - exocrine portion, acinar/duct cells, aqueous/enzymatic components of pancreatic juice
   fcns of: HCO_3^-, amylase, lipase, carboxypeptidase, trypsin, chymotrypsin (zymogenic forms -
   enterokinase)
large intestine = colon; bacterial metabolism - vit K, flatus; water/salt absorption; mass movement
KNOW   T24-2 (fcn of GI secretions), T24-3 (fcn of GI hormones), T24-4 (digestion summary - know
   appropriate enzymes, endproducts, locations)
absorption of nutrients - aa & sugars - facilitated diffusion/cotransport; lipids - FA + bile salts →
   micelles → reform triglycerides, coat with protein → chylomicron → lacteal

Chapter 25

nutrition - nutrients (essential) - CHO, lipids, proteins, vitamins, minerals, water
metabolism - anabolism/catabolism
Calorie/kilocalorie
food pyramid - interpretation
major nutrients - recommended dietary intake, usage in body
  CHO (simple sugars, starch, cellulose [bulk]), 1° fuel source, 4 cal/g, 125-175 g/day, complex better
  lipids - triglycerides (saturated/unsaturated), >30% caloric intake, limit satd FA & cholesterol,
    essential FA (linoleic acid - used to make PGs), 9 cal/g
  proteins - essential/nonessential aa, complete/incomplete proteins, ~12% caloric intake, N balance,
    4 cal/g
  vitamins - coenzymes, water-soluble (C & B complex), fat-soluble (A,D,E,K), hypervitaminosis
  minerals - cofactors, eg. Fe/Ca/P
interconversion of nutrients (liver) - glycogenesis, lipogenesis, glycogenolysis, gluconeogenesis
catabolism of sugars - glycolysis - w/in cytoplasm, no O_2 requirement, glucose → pyruvate;
  preparation + Krebs cycle - w/in mitochondrial matrix, pyruvate → acetyl CoA → CO_2;
  ETS - inner mitochondrial membrane, O_2 needed, $NADH/FADH_2$ → ATP
catabolism of proteins - oxidative deamination (rare); catabolism of fats - β-oxidation
metabolic rate - energy produced & used/time
3500 C = 1 lb body fat; Cal for maintenance = females (wt x 12), males (wt x 11)

Review Sheet - Chapters 26,27,28

Chapter 26

fcns of urinary system
kidney - renal cortex, renal columns, renal pyramids (medulla)
nephron = functional unit, cortical vs. juxtamedullary nephrons
glomerulus → Bowman's capsule → PCT → descending limb (loop of Henle) → ascending limb
  (loop of Henle) → DCT → collecting duct → renal papilla → minor calyx → major calyx → pelvis → ureter → urinary bladder → urethra
renal a → segmental a → interlobar a → arcuate a → interlobular a → afferent arteriole →
  glomerulus → efferent arteriole → peritubular capillaries → interlobular v → arcuate v →
  interlobar v → renal v
renal corpuscle = glomerulus + Bowman's capsule
juxtaglomerular apparatus = JG cells (w/in afferent arteriole, release renin) + macula densa (DCT)
filtration membrane, filtration slits, podocytes (BC)
urine production - filtration + reabsorption + secretion

Filtration - plasma filtered w/in glomerulus, filtrate enters Bowman's capsule
GFP = GCP (60 mmHg) - BCOP (32 mmHg) - COP (18 mmHg) = 10 mmHg
  importance of hydrostatic vs. osmotic forces involved; anything under 7 nm/40,000 daltons is filtered
GFR = 125 ml/min; = 180 L/day; 99% reabsorbed, 1% excreted as urine

Reabsorption - occurs along length of kidney tubules, T 26-3
PCT - proteins by endocytosis; active (or cotransport) transport - aa, glucose, Na/K/Cl/Mg/Ca/PO$_4$ ions
  - water moves by osmosis (reduces filtrate by 65%)
Loop of Henle - descending limb - H$_2$O permeable (reduces filtrate by another 15%);
  -ascending limb - H$_2$O impermeable; active/co-transport of Na/Cl/K ions
DCT - active transport of Na/Cl/Ca ions; water movt, reduces filtrate another 10%
CT - water movt, reduces filtrate another 9%   [DCT & CD water movt influenced by ADH]

Secretion - movt out of peritubular capillaries & into renal tubules for excretion; eg, H/K ions, drugs, etc

Urine concentration mechanism - countercurrent flow in Loop of Henle & vasa recta, urea movt
  Fig. 26.16; Osm varies from cortex → medulla, 300 → 1200 mOsm
  - in vasa recta - H$_2$O moves out descending limb & into ascending limb; salt moves out of
         ascending limb & into descending limb
  - in Loop of Henle - H$_2$O moves out of descending limb; salt moves out of ascending limb
  - urea - moves out of CD & into descending limb of Henle

Regulation of urine formation -
aldosterone → Na$^+$/H$_2$O reabsorption (K$^+$ secretion) → dec. urine output, inc. blood volume/BP
renin → angiotensin II → aldosterone release, vasoconstriction, thirst
ADH → H$_2$O reabsorption (DCT,CD) → dec. urine output, inc. blood volume/BP
ANP → dec. ADH release → inc. urine output, dec. blood volume/BP
autoregulation - w/ inc. BP → afferent arteriole constricts → dec. blood flow;
  w/ dec. BP → afferent arteriole dilates → inc. blood flow
sympathetic innervation - SNS causes vasoconstriction of afferent arterioles

pathologies/treatments - dialysis, calculi, diuretics (caffeine, EtOH)
micturition reflex - internal (smooth)/external (skeletal) urinary sphincter

Chapter 27

intracellular/extracellular (plasma & interstitial fluid) fluid compartments
water balance: intake(food + metabolic) = elimination (urine, feces, evaporation - lungs, skin); ADH
Na^+ balance - 1° extracellular cation, aldosterone regulation; hypo-/hypernatremia (T27-6)
K^+ balance - 1° intracellular cation, aldosterone regulation; hypokalemia → hyperpolarization;
 hyperkalemia → hypopolarization   (T27-7)
Cl^- balance - 1° extracellular anion, passively follows Na^+ movt
Ca^{+2} balance - PTH elevates/calcitonin decreases blood Ca^{+2} levels (T27-8)

Acid-Base balance - buffers: H_2CO_3/HCO_3^- (extracellular); $H_2PO_4^-/HPO_4^{-2}$ (intracellular);
 NH_4^+/NH_3 (in kidneys) ; proteins (intra- & extracellular usage)
 - on left of / : species that acts as an acid (proton donor), neutralizes base
 - on right of / : species that acts as a base (proton acceptor), neutralizes acid
respiratory mechanism - w/ high pH → hypoventilation → dec. pH; w/low pH → hyperventilation →
 get rid of excess CO_2 → inc. pH
renal mechanism - H^+/HCO_3^- secretion &/or reabsorption
acidosis vs. alkalosis - respiratory vs. metabolic causes (Table A, p. 1009)

Chapter 28

seminiferous tubules → rete testis → vas efferentia → epididymis → vas deferens →(inguinal canal) →
 ejaculatory duct → prostatic urethra → membranous urethra → spongy urethra
spermatogonium (1N) → 1° spermatocyte → 2° spermatocyte (1N) → spermatids → spermatozoa
 4 haploid sperm/spermatogonium
seminiferous tubules - produce sperm; Sertoli (nurse) cells - blood/testes barrier
epididymis - sperm storage/maturation; vas deferens - convey sperm into pelvic cavity
ejaculatory duct = vas deferens + duct draining seminal vesicle
semen (2-5 ml) = sperm (5%) + seminal vesicles (60%, fructose, PGs, fibrinogen) + prostate gland
 (30%, alkaline, clotting factors) + Cowper's gland (5%, alkaline, preejaculate)
fcn of coagulation of semen, fcn of testes placement w/in scrotum
emission, ejaculation, resolution; impotence, male infertility
vascular action associated with flaccid → erect penis; corpora cavernosa
GnRH, FSH (bind to Sertoli cells in seminiferous tubules), LH (binds to Leydig cells)
testosterone - produced by Leydig cells, functions

oogonium → 1° oocyte (arrested in Pro I, + granulosa cells = 1° follicle) → 2° oocyte (arrested in Meta II,
 matures in graafian follicle) → maturation to ovum only with fertilization
day 1-5 = menses; day 5-14 = follicular/proliferative phase; day 14 = ovulation; day 15-28 = luteal/
 secretory phase
ovarian cycle: *****Fig. 28.17 & T28-2*****
GnRH → inc. FSH/LH production; surge in LH/FSH @ day 14; FSH/LH → follicle development;
 LH surge → ovulation, ruptured follicle → corpus luteum (w/o pregnancy becomes corpus albicans)
 estrogen production begins proliferation of endometrium
HCG - with pregnancy, maintains corpus luteum (source of estrogen/progesterone prior to placenta)
uterine cycle:
changes in uterine lining due to estrogen/progesterone; ready for implantation by day 21; degenerates
 is pregnancy doesn't occur; progesterone "maintains" pregnancy - prevents myometrium contractions
PMS - days 25-28, due to dec. in steroid hormones
egg viable - 24 hrs; sperm viable 48-72 hrs; fertile period ~ days 11-15
fertilization w/in top 1/3 of fallopian tubes; implantation of blastocyst @ day 21 (trophoblast action)
zygote → blastocyst → gastrula (germinal period → 2 wks)→ embryo (→ 8 wks) → fetus (→ 9 mos)
control of pregnancy - sterilization, chemical/mechanical methods, prevent implantation
lactation - estrogen, prolactin, oxytocin; menopause